SpringerBriefs in Applied Sciences and Technology

Manufacturing and Surface Engineering

Series editor

Joao Paulo Davim, Aveiro, Portugal

More information about this series at http://www.springer.com/series/10623

Kush Mehta · Kapil Gupta

Fabrication and Processing of Shape Memory Alloys

 Springer

Kush Mehta
Department of Mechanical Engineering,
    School of Technology
Pandit Deendayal Petroleum University
Gandhinagar, India

Kapil Gupta
Department of Mechanical and Industrial
    Engineering Technology
University of Johannesburg
Johannesburg, South Africa

ISSN 2191-530X   ISSN 2191-5318   (electronic)
SpringerBriefs in Applied Sciences and Technology
ISSN 2365-8223   ISSN 2365-8231   (electronic)
Manufacturing and Surface Engineering
ISBN 978-3-319-99306-5   ISBN 978-3-319-99307-2   (eBook)
https://doi.org/10.1007/978-3-319-99307-2

Library of Congress Control Number: 2018953643

This Springer imprint is published by the registered company Springer Nature Switzerland AG
The registered company address is: Gewerbestrasse 11, 6330 Cham, Switzerland

Preface

The "smart" materials like shape memory alloys (SMAs) are a preferable choice for various aerospace, biomedical, automotive, industrial, and scientific applications due to their superior properties such as biocompatibility, light weightiness, robustness, and special properties in actuation, vibration, damping, and sensing. Since their inception to till date, there have been continuous research and developments efforts towards making fabrication and processing of these alloys easier. The main objective of this book is to disseminate research and developments as regards to the fabrication and processing of SMAs, and to facilitate specialist, engineers, and scientists working in the field with an aim to encourage the researchers to establish the field further.

This book consists of four chapters. It starts with Chap. 1 as an introduction to shape memory alloys where the history, types, and properties of SMAs as regards to their unique applications are highlighted. Chapter 2 sheds light on conventional and advanced machining of SMAs. As regards to the advanced machining, it majorly incorporates research aspects of machining SMAs by electric discharge machining, laser beam machining, and abrasive water jet machining. Whereas, the other part of this chapter is focused on difficulties encountered in conventional machining, and novel techniques and sustainable strategies to enhance the machinability of SMAs. Chapter 3 addresses different welding and joining processes attempted for SMAs. Various welding and joining processes such as tungsten inert gas welding, plasma welding, laser beam welding, electron beam welding, resistance welding, friction stir welding, friction welding, explosive welding, ultrasonic welding, diffusion bonding, adhesive bonding, brazing, and soldering are discussed from the point of view of process capabilities and challenges to fabricate SMAs. Chapter 4 presents processing techniques such as powder technology, additive processing, thermo-mechanical processing, and mechanical processing for SMAs. It covers the description of individual process principle, process capabilities, process parameters, and properties obtained after fabricating SMAs. The information presented in this book is majorly from the research conducted in this area.

Gandhinagar, India   Kush Mehta
Johannesburg, South Africa   Kapil Gupta

Contents

Chapter 1
Introduction

1.1 Introduction

Shape memory alloys (SMAs) or "smart alloys" are a unique class of smart materials that can change their form (shape or size), and can return back to their original form with applied heat, stress, or magnetic field. They have the ability to produce very high actuation strain, stress, and work output due to reversible martensitic phase transformations [1–4]. Shape memory alloys are compact, robust, lightweight, frictionless, quiet, biocompatible, environmentally friendly, and possess superior properties in actuation, vibration damping, and sensing.

Shape memory alloy was first discovered by a Swedish physicist Arne Ölander in 1932 [5]. Two researchers of the US Naval ordnance laboratory, William Buehler, and Frederick Wang discovered the shape memory effect (SME) in a nickel–titanium (NiTi) alloy in 1962 [6]. In 1970s, shape memory alloys were started using in the commercial products and devices. Since then, there has been tremendous developments in the field.

Due to the aforementioned unique properties, the demand for SMAs has been increasing for the applications, such as micro-electromechanical systems (MEMS); robotics; consumer and industrial products; automotive, aerospace, and biomedical fields. Figure 1.1 presents the use of SMAs in some specific application areas.

Shape memory alloys are used in actuators for automotive and aerospace applications. Along with actuators, structural connectors, vibration dampers, manipulators, and some pathfinder applications in aerospace are also deploying shape memory alloys. SMAs have also been successfully used in micro-and macro-actuators for robotic fingers and hand, and artificial muscles for robotics applications [3]. SMA-based actuators have been used in controlling flying robots. NiTi shape memory alloy is a prime candidate material for biomedical applications and used to make surgical tools, dental implants, bone implants, stents, and other medical equipment.

While manufacturing parts/components of SMAs to be used in the field stated above, these materials have to undergo extensive fabrication and processing opera-

© The Author(s), under exclusive licence to Springer Nature Switzerland AG 2019
K. Mehta and K. Gupta, *Fabrication and Processing of Shape Memory Alloys*,
Manufacturing and Surface Engineering, https://doi.org/10.1007/978-3-319-99307-2_1

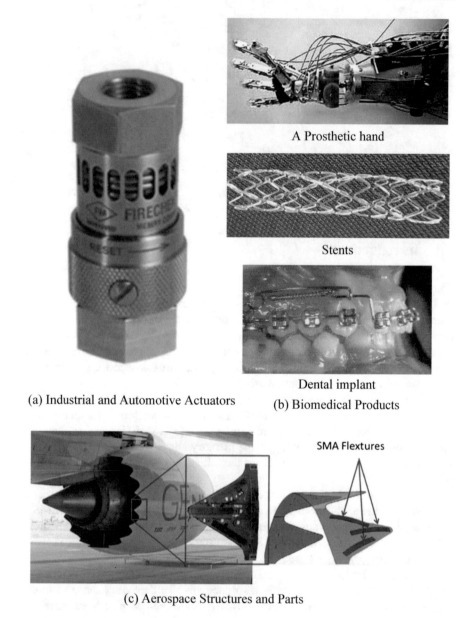

A Prosthetic hand

Stents

Dental implant

(a) Industrial and Automotive Actuators   (b) Biomedical Products

SMA Flextures

(c) Aerospace Structures and Parts

Fig. 1.1 Applications of shape memory alloys [3], with kind permission from Elsevier

tions. Practically, SMAs can exist in two different phases with three different crystal structures, and six possible transformations [3]. While heating, the SMA begins to transform from martensite into austenite phase. The two temperature points during this transformation are—"A_s-austenite-start-temperature", where transformation starts and 'A_f-austenite-finish-temperature', where transformation ends. Once an SMA is heated beyond A_s it begins to contract and transform into the austenite structure, i.e., to recover into its original form. This transformation is possible even under high applied loads, and therefore, results in high actuation energy densities [3]. During the cooling process, the transformation starts to revert to the martensite at martensite-start-temperature (M_s) and is complete when it reaches the martensite-finish-temperature (M_f). The highest temperature at which martensite can no longer be stress induced is called M_d, and above this temperature, the SMA is permanently deformed like any ordinary metallic material [3]. Three possible shape change effects in SMAs are given in the subsequent sections.

1.1.1 Pseudoelasticity

Due to pseudoelasticity (also called superelasticity), the SMA reverts to its original shape when the applied deformation stress is removed [4]. It is done by the subsequent recovery of the deformation strain when the stress is removed. Figure 1.2 illustrates the pseudoelasticity mechanism where the material from point "A" is stressed at a constant temperature while being in a stable austenite phase. The resulting deformation is elastic until a certain point "B", where the material reaches the state that the martensitic transformation begins. From this point on, the transformation that takes place is accomplished under a constant stress, while the strain continues to increase, until a maximum strain level (point C). The maximum strain level varies according to the material. The curved section between points (B) and (C) is termed the stress "plateau." At this point, the phase transformation from austenite to martensite is completed and the curve that describes the behavior of the material is different. This new behavior usually presents a small temperature hysteresis ($\Delta T = A_f - M_s$), whereas the "parent" martensite interfaces present some mobility. Further stressing the material from point (C) will only lead to elastic deformations of the detwinned martensite. Finally, after the stress is removed, the material begins to return to its stable austenite phase, until it fully transforms this phase (point D), thus the cycle can be repeated.

1.1.2 Shape Memory Effect

The one-way shape memory effect retains a deformed state after the removal of an external force and then recovers to its original shape upon heating. The two-way or reversible shape memory effect can remember its shape at both high and low temperature, but it is less applied commercially. Shape memory effect in terms of microstructure and mechanism is shown in Fig. 1.3. SMA is having twinned martensitic structure at low temperature can be subjected to detwinned microstructure as shown in Fig. 1.3a by reorientation of martensitic grains. This detwinning retains the

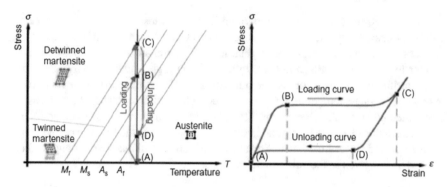

Fig. 1.2 Stress–temperature (left) and stress–strain (right) curves that describe the superelasticity behavior [4], with kind permission from Elsevier

shape of material after releasing load as it affects macroscopic shape change. Consequently, the heating of SMA leads to reverse phase transformation when heated above A_f temperature such as detwinned martensitic structure to the austenitic structure that subsequently completes material shape recovery as shown in Fig. 1.3b. Further, cooling SMA below M_f causes formation of twinned martensitic structure without shape change. This is called shape memory effect. The detwinning initiation caused by minimum stress is shown as the detwinning start stress (σ_s) while the detwinning finish stress (σ_f) shown for complete detwinning of martensitic structure resulted from sufficiently higher load [2].

1.1.3 Transformation-Induced Fatigue

SMA is having unique characteristics of shape memory effect, psudoelasticity, and thermal phase transformation. Repetition of the abovementioned cyclic process of thermo-mechanical analysis can lead to the permanent deformation under applied loading condition after certain number of cycles and called as premature fatigue of SMA. This is dependent on different factors such as various fabrication methods (for example, additive manufacturing, powder metallurgy, welding, heat treatment, forming, and casting, etc.), environmental conditions (for example, humidity, and temperature, etc.), microstructural modifications at the time of transformation and working conditions (like stress, and strain, etc.) [2, 4]. The microstructural changes are expected with cyclic loading considering mechanical and thermal changes. Subsequently, the shape memory behavior is deteriorated due to this microstructural change occurred mechanically as well as thermally [2, 3].

Fatigue life can be calculated based on the number of cycles performed depending on magnitude of applied stress within elastic region, wherein the phase transformation may be mechanically completed such as complete martensitic to austenitic transformation or partially completed such as start or end point is at the mid of specific

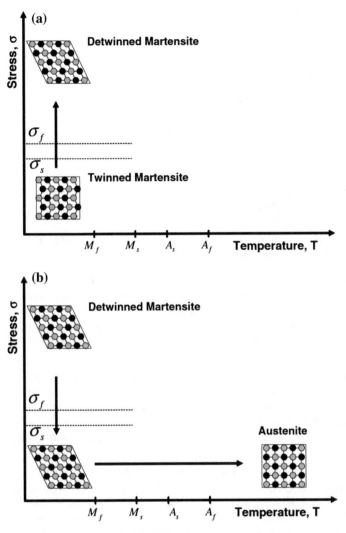

Fig. 1.3 Shape memory effect, **a** material detwinning with an applied stress and **b** unloading and heating to austenite [2]

structure. In case of thermal loading, the fatigue life is calculated based on transformation strain occurred during each cycle (similar as mentioned earlier as complete or partial). Further, crystallographic orientation of grains and formation of precipitates in microstructure are also responsible for fatigue life of SMAs. Additionally, the oxidation and corrosion lead to the fast crystal degradation and subsequently affect fatigue life. The heat treatment processes can also be performed to vary the fatigue life of the SMAs [2].

1.2 Types of Shape Memory Alloys

There exists a wide range of metals, polymers, ceramics, and other materials that show the shape memory effects. Shape memory alloys (metals) are available in a wide range, as given below [3, 4]:

- Iron-based alloy Fe–Mn–Si;
- Copper-based SMAs such as Cu–Zn–Al; Cu–Al–Ni, Cu–Al–Ni–Mn, and Cu–Sn;
- Nickel–Titanium based alloys such as NiTi, NiTiCu, NiTiPd, NiTiFe, NiTiNb, NiFeGa, and NiTiCo;
- Kovar (29% Ni, 17% Co, 0.3% Si, 0.1% C and Fe balance);
- Hi-temperature shape memory alloys such as TiNiPd, TiNiPt, NiTiHf, NiTiZr, ZrRh, ZrCu, ZrCuNiCo, ZrCuNiCoTi, TiMo, TiNb, TiTa, TiAu, UNb, TaRu, NbRu, and FeMnSi;
- Magnetic shape memory alloys, namely, NiMnGa, FePd, NiMnAl, FePt, Dy, Tb, LaSrCuO, ReCu, NiMnIn, and CoNiGa.

The NiTi alloys have greater shape memory strain (up to 8% vs. 4–5% for the copper and iron-based alloys). Although the NiTi alloys are expensive but possess excellent biocompatibility, and corrosion resistance, and much more thermally stable compared to the other copper based alloys. Therefore, the NiTi shape memory alloys find extensive biomedical applications. They are used to manufacture dental and orthopedic implants, cardiovascular stents, and surgical tools and instruments. Pipe joints for steel pipes and fishplates for crane rails etc. are some important application areas where iron-based SMAs are used. Copper-based SMAs possess higher actuation temperature are, therefore, used in high-temperature applications, and in actuators for several automotive, robotics, and industrial applications.

Despite the aforementioned characteristics and important applications of SMAs, they are considered as difficult-to-fabricate and process materials because high ductility, typical stress–strain behavior, low thermal conductivity, and high degree of work hardening of these alloys lead to poor chip breaking, burr formation, and progressive tool wear while machining, occurrence of weld defects and low joint strength during their welding and joining, nonuniform melting and defects in part microstructure at the time of other processing operations [7–14]. These all consequently results in poor work surface integrity, very high consumption of energy and resources, escalated cost of fabrication and processing and high environmental footprints.

To overcome the aforementioned challenges, as regards to the fabrication and processing of SMAs, significant attempts, i.e., innovations, and research and developments have been made to improve part quality, reduce cost, and enhanced safety and sustainability.

The forthcoming chapters of this book shed light on various important aspects of fabrication and processing technologies such as conventional, advanced, and sustainable machining; welding and joining; additive manufacturing; forging and rolling; and powder metallurgy of shape memory alloys.

References

1. K. Otsuka, C.M. Wayman (eds.), *Shape memory materials* (Cambridge University Press, Cambridge, 1999)
2. P.K. Kumar, D. Lagoudas, Introduction to shape memory alloys. in *Shape Memory Alloys* (Springer, Boston, MA, 2008) pp. 1–51
3. J.M. Jani, M. Leary, A. Subic, M.A. Gibson, A review of shape memory alloy research, applications and opportunities. Mater. Des. **56**, 1078–1113 (2014)
4. A.P. Markopoulos, I.S. Pressas, D.E. Manolakos, Manufacturing processes of shape memory alloys. in Materials Forming and Machining (Research and Development, 2016) p. 155
5. A. Ölander, An electrochemical investigation of solid cadmium-gold alloys. J. Am. Chem. Soc. **54**(10), 3819–3833 (1932)
6. W.J. Buehler, J.V. Gilfrich, R.C. Wiley, Effect of low-temperature phase changes on the mechanical properties of alloys near composition TiNi. J. Appl. Phys. **34**(5), 1475–1477 (1963)
7. Y. Guo, A. Klink, C. Fu, J. Snyder, Machinability and surface integrity of Nitinol shape memory alloy. CIRP Ann. Manuf. Technol. **62**(1), 83–86 (2013)
8. M.H. Sadati, Y. Javadi, Investigation of mechanical properties in welding of shape memory alloys. Procedia Eng. **149**, 438–447 (2016)
9. O. Akselsen, Joining of shape memory alloys, in *Shape Memory Alloys*, ed. by O. Akselsen (InTech, USA, 2010)
10. Y. Wang, S. Jiang, Y. Zhang, Processing map of NiTiNb Shape memory alloy subjected to plastic deformation at high temperatures. Metals **7**(9), 328 (2017)
11. M.H. Elahinia, M. Hashemi, M. Tabesh, S.B. Bhaduri, Manufacturing and processing of NiTi implants: a review. Prog. Mater Sci. **57**(5), 911–946 (2012)
12. J.P. Oliveira, R.M. Miranda, F.B. Fernandes, Welding and joining of NiTi shape memory alloys: a review. Progress in Materials Science **88**, 412–466 (2017)
13. M. Elahinia, N.S. Moghaddam, M.T. Andani, A. Amerinatanzi, B.A. Bimber, R.F. Hamilton, Fabrication of NiTi through additive manufacturing: a review. Prog. Mater Sci. **83**, 630–663 (2016)
14. Y. Zhou, Micro-welding of shape-memory alloys. in *Joining and Assembly of Medical Materials and Devices* (2013) pp. 133–153

Chapter 2
Machining of Shape Memory Alloys

2.1 Introduction

As discussed in Chap. 1, the shape memory alloys (SMAs) are considered as difficult-to-machine (DTM) materials. In other words, SMAs possess poor machinability that subsequently lead to the problems such as progressive tool wear, deterioration of surface quality, burr formation, and high consumption of energy and resources, which are encountered during their machining [1–3]. For example, Fig. 2.1 shows some drawbacks while machining Nitinol (NiTi) SMA during conventional grinding and turning processes.

Manufacturing sector strives for the technological development as regards to facilitating the machining of DTM materials including SMAs due to the lack of available data for their machining. Two important machinability aspects such as better quality and reduced cost are the prime factors that drive the research and development in the manufacturing sector. The following strategies and techniques could be adopted to obtain enhanced machinability in DTM materials [4, 5]:

- Employing advanced (electric discharge, laser, and abrasive water jet machining) and hybrid machining processes (heat- and vibration-assisted machining, etc.) in order to reduce the process chain by eliminating the need of subsequent finishing operations;
- Selection of optimum machining conditions to minimize energy consumption and to maintain cost efficiency;
- Selecting suitable tool materials-geometries-coatings to minimize tool failure and improving part quality thereby maintaining resource efficiency;
- Adopting advanced lubrication/cooling techniques such as minimum quantity lubrication (MQL), and cryogenic cooling, etc. in order to minimize tool wear rate, improve work surface quality, maximize productivity, and cut down the cost associated with the use of lubricant and lubrication system with extremely low environmental impact.

© The Author(s), under exclusive licence to Springer Nature Switzerland AG 2019
K. Mehta and K. Gupta, *Fabrication and Processing of Shape Memory Alloys*,
Manufacturing and Surface Engineering, https://doi.org/10.1007/978-3-319-99307-2_2

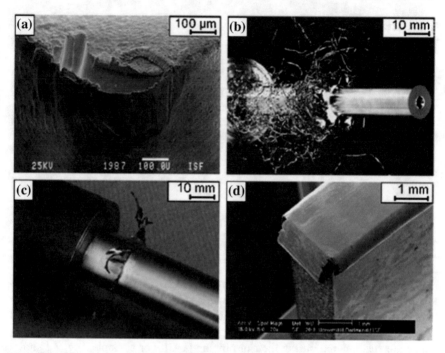

Fig. 2.1 Problems while machining of Nitanol (NiTi) shape memory alloys **a** high tool wear; **b** adverse chip form; **c** formation of burrs after turning **d** and grinding [3]

Using the aforementioned strategies and corresponding techniques, several attempts have been made to improve the machinability of SMAs. The following sections comprehensively discuss the machinability behavior of SMAs under different machining environments.

2.2 Advanced Machining of Shape Memory Alloys

2.2.1 Electric Discharge Machining

Electric discharge machining (EDM) also known as spark erosion machining is a thermal type advanced machining process, where the material removal takes place by melting and vaporization caused due to series of repeated electrical discharges. These electric discharges occur between electrode cathode (tool) and anode (workpiece) under suitable dielectric fluid specifically in the inter-electrode gap when a pulsed DC power is supplied (see Fig. 2.2) [6]. After its inception in 1878, EDM was started using for commercial machining purposes in 1930. Electric discharge machining has been the first choice of manufacturers for near-net-shape manufacturing of dies and

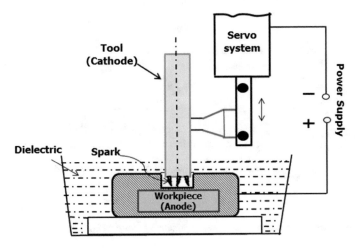

Fig. 2.2 Schematic representation of working principle of electric discharge machining

molds; micro parts and features; machining of superalloys, bio-materials, and other difficult-to-machine materials for automobiles, aerospace, nuclear, and biomedical applications [6, 7].

Wire electric discharge machining (WEDM or wire-EDM) is an important and most extensively used variant of EDM where a continuously moving thin wire (acts as cathode) is used as a tool to cut or machine the workpiece (anode) under a continuously flowing dielectric (see Fig. 2.3) [8]. Pulse-on time, pulse-off time, voltage, current, tool electrode feed rate, and dielectric type, etc., are the important parameters of EDM and WEDM. There has been a past track record to machine difficult-to-machine materials prominently by these processes. Optimal parameter settings, hybridization, and using sustainable techniques, etc., are the key factors achieving high surface integrity aspects such as high geometric accuracy, fine surface finish, favorable microstructure, and good tribology and wear characteristics while machining DTM materials including SMAs by EDM and WEDM processes.

A detailed study on EDM of NiTi with Cu electrode has been done where discharge energy mode with pulse current and duration is identified as the key parameter for precise machining [9]. Longer pulse duration and lower pulse current are identified for better material removal rate and lower recast layer. A similar study on EDM of NiAlFe reveals the importance of energy and recommends low energy settings for minimum recast layer thickness and maximum material removal rate [10]. Effects of two most important EDM parameters mainly discharge current and pulse duration on characteristics of $Ni_{60}Al_{24.5}Fe_{15.5}$ have been investigated in detail. Figure 2.4 depicts variation of material removal rate, surface roughness, and electrode wear rate with pulse duration and current. It is seen that high material removal rate due to more material melting, evaporation and high impact force of expanded dielectric medium is observed at high discharge current that probably can have high current density. Deterioration in surface roughness has been observed with increased discharge cur-

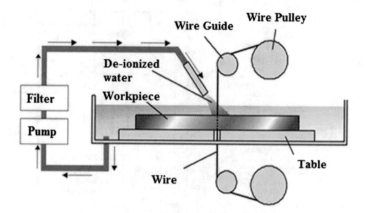

Fig. 2.3 Schematic representation of working principle of wire-EDM

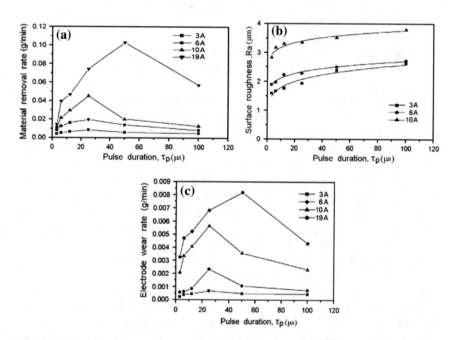

Fig. 2.4 Effect of pulse duration on **a** material removal rate; **b** electrode wear rate; and **c** surface roughness during EDM of $Ni_{60}Al_{24.5}Fe_{15.5}$ alloy [10], with kind permission from Elsevier

rent and pulse duration. Worsened erosion effect due to large amount of discharge at high current and generation of larger and deeper craters at longer pulse duration are the main reasons reported [10].

Abidi et al. [11] conducted an investigation on the effects of micro-EDM parameters, namely, capacitance, voltage, and electrode material on machinability of NiTi SMA. Multi-performance optimization of micro-EDM parameters secures their best values (475 pF, 80 V, and use of brass electrode) for good geometric accuracy, less overcut, and good surface finish of the machined hole. A systematic study on the influence of EDM parameters and their optimization during NiTi machining concludes the positive effect of pulse current and pulse-on time on material removal rate, but alongside it deteriorates the surface quality and reduces tool electrode life, hence a trade-off in EDM parameters is identified that requires multi-performance optimization [12, 13].

Powder mixed rotary EDM was also explored and found very effective in tool wear reduction and productivity enhancement [14]. Powder mixed EDM with optimum combination of machining parameters (especially tool rotational speed) manages the ion's movement, spark intensity, and penetration, and thus reduces the tool wear, increases the material removal rate, and decreases the surface roughness.

A researcher named *Mallaiah M* has conducted a systematic and detailed investigation on Wire-EDM of NiTi shape memory alloys [15–19]. The material removal rate and surface roughness while WEDM of $Ti_{50}Ni_{40}Cu_{10}$ shape memory alloy have been investigated [15]. Increasing material removal rate with pulse-on and improving surface roughness characteristics with pulse-off time have been found. High discharge energy and intensity of spark at long pulse-on time; and adequate flushing in the machining zone due to longer pulse-off time are the reasons behind that. Later, desirability based multi-performance optimization secured optimal WEDM parameters for the best machinability of SMA with 1.83 μm average roughness and 7.6 mm^3/min material removal rate. Furthermore, the XRD study informed the presence of thermal damage as the formation of oxide and other phases took place corresponding to longer pulse-on time [16]. Low pulse-on time with high voltage has been recommended for enhanced surface quality. Another investigation on $Ti_{50}Ni_{40}Cu_{20}$ also reports almost same trends and additionally with higher hardness along longer pulse-on time due to high discharge melting and rapid solidification of molten material that probably increased the carbide content percentage on the machined surface of SMA [17]. Interaction between pulse-on time and voltage has been identified as one of the most important factors to be considered in WEDM of SMAs. Formation of craters, pockmarks, pits, and globules due to increased surface roughness at higher peak current and longer pulse-on time is shown in Fig. 2.5 [18].

SMA machining with coated brass wire resulted in low MRR compared to plain brass wire [19]. Interestingly, machining with increased wire speed caused higher material removal rate as the molten material is splashed around the surface by flushing pressure and discharge of large gas volume from the molten pool increases material removal rate [19].

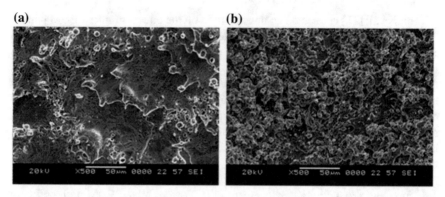

Fig. 2.5 SEM micrographs of the machined surface of $Ti_{50}Ni_{40}Cu_{10}$ SMA **a** at 2A peak current **b** at 12 A peak current [18], with kind permission from Elsevier

2.2.2 Laser Beam and Abrasive Water Jet Machining

The principle concept of *LASER* (light amplification by stimulated emission of radiation) was originated by Albert Einstein in 1917. Thereafter, in 1960, Theodore H. Maiman built the first working LASER in the United States [20]. Since then, there has been a continuous development in the field of laser technology that facilitates its use in various areas of science, engineering, medicine, food processing, and various other activities.

Figure 2.6 illustrates the working principle of laser beam machining (LBM). It is a process where a high-intensity laser beam is used to rapidly heat the target work surface, subsequently melts and/or vaporizes the target material through the full depth to cut the required geometry or shape or to remove the required amount of material [20, 21]. The "Laser" as a massless tool is not subjected to wear and tear, and offers great flexibility. Carbon dioxide lasers, Nd: YAG lasers, fiber, and ultra-short pulse (femtosecond) lasers are the most common lasers used in machining applications. The important laser parameters such as laser power, pulse energy, scan speed frequency, focal distance, and gas pressure, etc., influence part quality significantly [21].

Like machining of other difficult-to-machine materials, shaping of intecrate features, and production of micro to macro parts; shape memory alloys have also been precisely cut by laser beam machining process. Nd: YAG laser machining of NiTi alloy is required to be done on low pulse energy setting for high geometric accuracy and surface finish, moreover it is necessary to obtain the optimal value of nozzle distance (that determines the speed rate of the gas) to avoid the dross formation and kerf [22]. The cutting speed is controlled by the pulse energy and the pulse energy is by laser power. It is worth mentioning that the pulse energy can be altered by altering the peak power while keeping the pulse width constant. High cutting speed obtains a decrease in energy in the machining zone, and thus produces low kerf. On contrary, at high speed, intense removal of material due to high energy input results in high roughness and kerf.

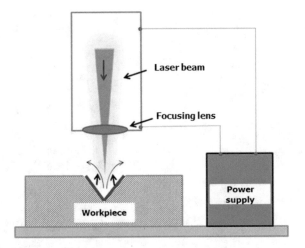

Fig. 2.6 Working principle of laser beam machining

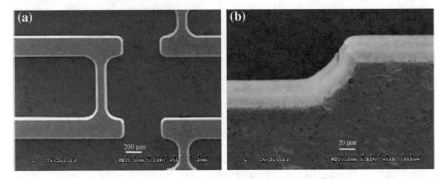

Fig. 2.7 A NiTi self-expanding medical micro-device fabricated by femtosecond laser machining **a** normal image **b** enlarged image [24], with kind permission from Elsevier

Micromachining of nickel–titanium tubes to be used in medical devices has also been done using femtosecond laser [23]. Various strategies on laser cutting path have been investigated and cutting in the zig-zag direction was found to be more beneficial that dramatically reduces heat-affected zone, debris accumulation, and machining time; and improves machining efficiency and cost-effective [23].

In a study of similar kind, high scanning/cutting speed for low thermal damage and smoother surface is recommended while laser machining of NiTi shape memory alloy for fabrication of precision miniature devices [24]. Using the obtained optimum value of scanning speed 60 mm/min and at 40 J/cm^2 of laser fluence, some parts of medical micro-device were fabricated (see Fig. 2.7). These parts are free from recast layer and possess high surface quality.

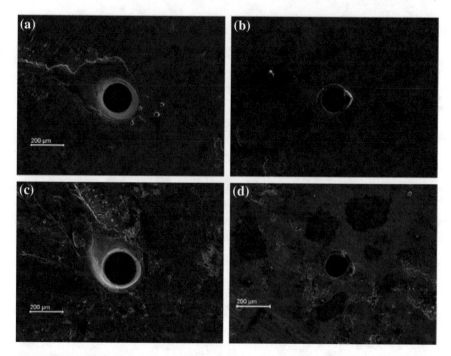

Fig. 2.8 SEM images of laser drilled holes in $Ni_{46}Mn_{27}Ga_{27}$ SMA at 500 W power (hole entrance at left and exit at right), pulse duration **a, b** 10 ms **c, d** 25 ms [25], with kind permission from Elsevier

Shape memory alloy have also been drilled precisely using fiber laser [25]. The holes drilled in $Ni_{46}Mn_{27}Ga_{27}$ ferromagnetic shape memory alloy at various levels of pulse duration and peak power. SEM images of the entrance and exit of the laser drilled holes are shown in Figs. 2.8 and 2.9.

It is observed that the increase in pulse duration increased the material removal rate as well as the diameter of the hole. Laser drilled holes are found to be of high geometric accuracy, i.e., excellent circularity.

Abrasive water jet machining is also one of the most extensively used advanced machining process. The high-speed water mixed with abrasive particles is used to remove materials from the workpiece in AWJM [26]. The kinetic energy of high-speed water jet is transferred to the abrasive particles and the mixture impinges on to the workpiece to remove the required amount of material or to form a particular shape/geometry. Water jet pressure; abrasive type, size and mass flow rate; and stand-off distance are the most important parameters of AWJM process. This process offers many significant advantages such as reduced wastage, no heat-affected zone, low environmental contamination, and no cutting fluid requirement, etc. Abrasive water jet machining process has been successfully employed to machine precision engineered parts and cut difficult-to-machine materials [26–28].

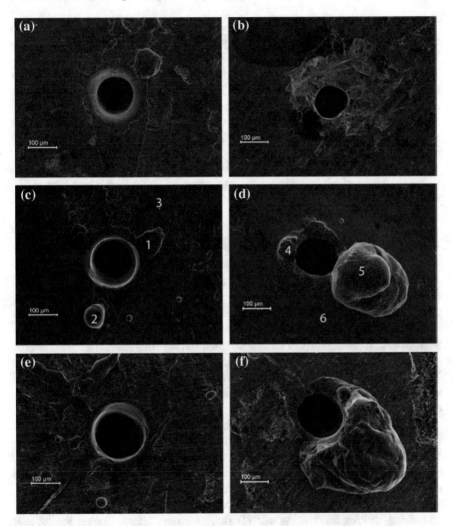

Fig. 2.9 SEM images of laser drilled holes in $Ni_{46}Mn_{27}Ga_{27}$ SMA at 1000 W power (hole entrance at left and exit at right), pulse duration **a, b** 1 ms **c, d** 10 ms, **e, f** 25 ms [25], with kind permission from Elsevier

Kong et al. [29] conducted plain and abrasive (garnet based) water jet milling of $Ni_{49.8}Ti_{50.2}$. Plain water jet milling on martensitic structure of NiTi resulted in excessive deformation and strain hardening that caused difficulty in controlling milling depth and flatness. The abrasive water jet milling was found more suitable in controlling depth of penetration and surface integrity aspects. Some grit embedment was observed that could easily be avoided by keeping the velocity low or by following the plain water jet milling as the successive post-processing technique.

Abrasive water jet milling of NiTi plates of different thickness has been conducted at various process parameters as water jet pressure- 70–345 MPa, traverse rate- 1–500 mm/min, abrasive mass flow rate-0.1–0.76 kg/min, and stand-off distance- 1–5 mm [30]. An optimal combination of 5 mm/min cutting speed and 345 MPa pressure, 0.26 kg/min mass flow rate with 3 mm stand-off distance has been obtained to precisely machine NiTi for low kerf widths. The magnitude to grit embedment has been found increasing from middle region to corner due to the dwell time existence and water jet lag. Least entrapment of particle/jet in the middle region probably reduced the grit embedment.

Overall, AWJM has been recommended for precise cutting/machining of shape memory alloys followed by water jet machining as a secondary finishing technique.

2.3 Sustainable Machining of Shape Memory Alloys

As discussed in the foregoing section that the advanced or nonconventional machining processes such as electric discharge machining, laser, and abrasive water jet cutting have successfully demonstrated their capability to enhance the machinability of SMAs. But considering the cost-effectiveness, specific geometric and dimensional requirements, quality of the end product, and ease of availability; conventional machining still plays a vital role to fabricate a range of components from SMAs. Machining at optimum parameters and utilizing effective tool materials, geometries and coatings, some successful attempts have been made to enhance the machinability of SMAs.

Selecting optimum machining parameters, Lin et al. [31] successfully improved the machinability of NiTi alloys during their drilling and mechanical cutting. The best drilling ability in terms of optimum values of cutting forces, machining time, and service life of the tool was observed with tungsten–carbide twist drill at 163 rpm rotational speed and 0.07 mm/rev feed rate.

A series of experimental work conducted by Weinert et al. [32–34] was focused on turning, drilling and milling of NiTi shape memory alloys. Cutting forces, tool wear and hardening of the machined subsurface zone were the important responses to evaluate the machining process. After turning of NiTi alloys at different cutting speeds (low–medium–high), Weinert and Petzoldt [3] concluded that 100 m/min cutting speed is best suited for optimum values of cutting forces, tool wear and burr formation to machine NiTi SMA. Moreover, low feed rates were analysed to form burrs and generate high tool wear, and hence strictly recommended not to be chosen. On contrary, for drilling of SMA medium cutting speed (30–60 m/min) with higher feed rates were found to produce the best results. Another experimental study conducted by Weinert et al. [32] recommends the use of coated (multilayer) carbide tools to minimize tool wear while turning NiTi alloy. By selecting the best machining practices such as the use of coated tools and machining at optimum parameters, etc. Weinert et al. successfully machined a pipe coupling of NiTiNb to connect hydraulic lines.

For the production of NiTi tube to be used in medical applications, it has to undergo extensive deep hole drilling operation. In an important study, using single-lip drills of cemented carbide material during deep hole drilling, NiTi tubing of high geometric accuracy and excellent surface quality have successfully been manufactured [34].

Sustainable machining strategies especially the usage of green lubricants and advanced cooling and lubrication techniques have been given considerable preference and have facilitated conventional machining with enhanced machinability. The combination of optimum parameters, tool coatings, etc., with sustainable techniques, have more pronounced effects on machinability.

Minimum quantity lubrication (MQL) is a micro-lubrication technique where small amount of green lubricant is supplied in the machining zone [4]. It facilitates near-dry machining and provides uniform lubrication at tool–chip interface rather than cooling. MQL technique manages the heat by reducing friction between chip and tool and thereby prevents heat to generate and protect tool and work surface from the adverse effects of high temperature. Furthermore, improves tool life, enhance work surface integrity, facilitates chip removal, and ensures safety and environment protection. In MQL, highly biodegradable and environmentally friendly green lubricants such as fatty acids, synthetic esters, vegetable oils etc. mixed with compressed air are supplied in the machining zone at flow rate ranges between 10 and 300 ml/h on contrary to high cutting fluid consumption (several liters per minute) in conventional flood cooling [35, 36].

Conventional wet cooling makes use of hydrocarbon-based harmful cutting fluids that may affect operator's health, causes high environmental footprints, and lead to high manufacturing cost. Moreover, the ineffective cooling effect results in extreme tool wear, deterioration of surface quality of the part being machined, and energy and resource inefficiency [35, 36].

Figure 2.10 illustrates an MQL system which consists of a cutting fluid/lubricant reservoir, a source of compressed air, nozzle and tubing. In flow control system, the atomization of cutting fluid takes place where it interacts with the highly compressed air. The atomized cutting fluid in the form of "micro-droplets" is sprayed into the machining zone through tubing and nozzle at a controlled flow rate and air pressure. Flow rate, air pressure, nozzle angle, and distance are some important parameters of MQL process.

Cryogenic cooling is another sustainable lubrication and cooling technique that uses cryogenic gases such as nitrogen, helium, and carbon dioxide as coolants for machining. These gases evaporate into the air and keep the environment clean and green [4]. Cryogenic gas cylinder, pressure gauge and flow control system, nozzle and tubing, as shown in Fig. 2.11 are the important components of a cryogenic cooling system. It works on the principle of spraying liquid nitrogen (at temp of approx. $-200\,^{\circ}C$) through a small diameter nozzle to the machining zone. While absorbing heat during machining, the liquid nitrogen forms a protective layer that covers tool surface and reduces tool–chip interface temperature. It increases tool life by reducing chemical reaction between tool and chip. The chips produced by this technique are easily recycled as no oil residue is attached.

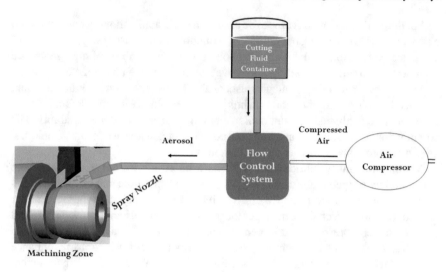

Fig. 2.10 Schematic of MQL system [4]

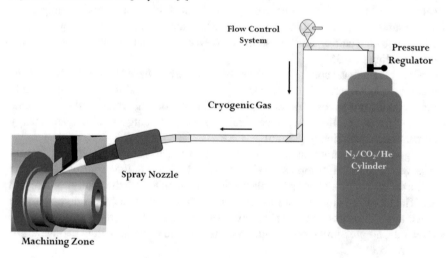

Fig. 2.11 Schematic representation of a cryogenic cooling system [4]

With regards to difficult-to-machine materials (including shape memory alloys) where high chemical reactivity at elevated temperature, low thermal conductivity, high hardness, and stress–strain rate are the major challenges, MQL is an effective technique that plays vital role to overcome these problems associated with their machinability. Past research endeavors have shown significant enhancement in machinability of shape memory alloys after employing sustainable cooling and lubrication techniques, i.e., MQL and cryogenic cooling.

Table 2.1 Summary of the past work on sustainable conventional machining of SMAs

Sl. no.	Researchers	Type of shape memory alloy	Process used
1	Weinert and Petzoldt [3]	NiTi (Nitanol)	Turning
2	Lin et al. [31]	NiTi	Drilling and mechanical cutting
3	Weinert et al. [32]	NiTi	Turning
4	Weinert and Petzoldt [33].	NiTi	Micro-milling
5	Weinert and Petzoldt [34]	NiTi	Deep hole drilling
6	Biermann et al. [37, 38]	NiTi	MQL-based Micro-milling
7	Piquard et al. [39, 40]	NiTi	Plain and MQL Micro-milling
8	Shyha et al. [41]	Kovar	Dry Drilling
9	Kaynak et al. [42–45].	NiTi	Turning (at dry, MQL, Cryogenic conditions)
10	Zailani and Mativenga [46]	NiTi	Micro-milling (at chilled air cooling and MQL conditions)
11	Kuppuswamy and Yui [47]	NiTi	High-Speed Micro-milling (at MQL with Johnson baby oil)
12	Akbari et al. [48]	NiTi	Turning (under the assistance of Ultrasonic vibrations)

The section below discusses various important aspects of the work conducted by researchers aimed to enhance machinability of shape memory alloys using sustainable machining techniques.

In a work based on green machining of shape memory alloys, minimum quantity lubrication (MQL) technique with ester oil was employed by Weinert and Petzoldt [33, 34]. Strong adhesive affinity of shape memory alloy to the cutting tool was the main reason behind the implementation of MQL. No NiTi adhesions were found to be reported after MQL based micro-milling with TiAlN coated solid carbide end-mill cutters. A high feed rate combined with high width of cut ensures extended tool life with least burr formation. The study facilitates the production of miniature parts of NiTi alloy by micro-milling.

Table 2.1 presents the exclusive summary of the past work (based on the extensive literature review) on conventional machining of shape memory alloys (SMAs).

A simulation-based experimental analysis of deep hole micro-drilling of SMAs has employed combined strategies of tool coatings and minimum quantity lubrication [37, 38]. Two-lip, PVD TiAlN-coated ball nose cutters of carbide with a diameter of 1 mm for milling, and single-lip drilling tools (both coated and uncoated) having diameters ranges from 0.5 to 1.5 mm and some twist drills were used. Some of the conclusions of their work are a combination of high width of cut and low cutting depth resulted in moderate tool wear, for minimum of ploughing and burr formation, cutter movement should be orthogonal to the chip thickness gradient and point in the direction of un-machined material, holes having diameters as small as 0.5 mm and a depth of up to 15 mm can be drilled with a high quality, twist drills performed better and allowed the drilling of NiTi SMA to much higher drilling depths and cutting speeds due to minimization of friction between the tool and hole wall as well as adhesion wear [37, 38].

Another investigation on burr formation and phase transformation characteristics during micro-end milling of NiTi alloy were done which resulted in the subsequent optimization of machining parameters to attain a significant improvement in machinability [39, 40]. Coated end mills of 0.8 mm diameter and of two teeth with 25° helix angle were employed for micro-milling. MQL technique at 6 bar air pressure and 1.6 ml/min flow rate for ester oil was also used. Experiments were designed and conducted based on half factorial design to the effects of depth of cut, width of cut, cutting speed, feed per tooth, strategy (up- and down-milling) on machinability of NiTi alloy. Lower values of feed per tooth and width of cut for smaller burrs and up-milling for thinner burrs were recommended. Optimization was done to facilitate the machining at high cutting speed with minimum burr formation.

Dry drilling of Kovar SMAs with HSS twist drills has been performed [41]. Experiments were conducted in dry condition at a speed range of 450–3750 rpm on unbacked and backed workpieces of Kovar alloy. Cutting speed was found to be the most significant factor affecting the surface integrity characteristics of Kovar. Lower feed rate and cutting speed together with smaller drill sizes are suggested to minimize the cutting temperature and thereby improving the surface integrity, i.e., reduction in thermal hardening and burr size and minimizing the chances of generation of micro-cracks, etc.

The research work on the use of advanced cooling and lubrication techniques to improve the machinability of SMAs was majorly contributed by Kaynak et al. [42–45]. A detailed and systematic investigation on the effects of dry machining, cryogenic cooling and MQL conditions on the machinability (specially tool wear and surface quality) of NiTi shape memory alloys was done. In a series of experimental work, Kaynak et al. found cryogenic machining as a promising approach for improving the machining performance of NiTi SMAs [42–45]. Effects of dry, cryogenic, preheated, and MQL conditions on machining characteristics were evaluated and analysed comprehensively. Liquid nitrogen at 1.5 MPa was used as cryogenic coolant, whereas 175 °C was chosen as preheating temperature for preheated machining, and 60 ml/h as flow rate and 0.4 MPa as air pressure for MQL at different combinations of machining parameters. One of their articles [42] reports on turning under dry, preheated and cryogenic machining conditions at three levels of cutting speed (12.5–25–50 m/min).

Figure 2.12 depicts cryogenic cooling arrangement made by Kaynak et al. for cutting tool while turning SMA.

As shown in Fig. 2.13, the tool wear study reveals a significant increase in tool wear at nose region in case of dry and preheated conditions compared with cryogenic machining with an increase in cutting speed. At high cutting speeds, cryogenic machining significantly reduced the notch wear as well as eliminated chipping (flaking) and chip flow damage. Furthermore, the cutting force decreased with increasing cutting speed under the influence of cryogenic machining condition (Fig. 2.14).

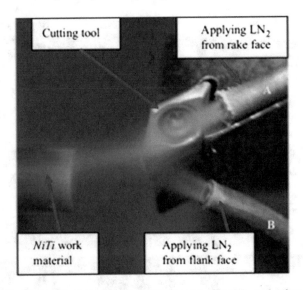

Fig. 2.12 The liquid nitrogen delivery system developed by Kaynak et al. where one nozzle is placed at the tool rake face and another one at the flank face (nozzle diameters are 4.78 mm; nozzle B makes 55° angle with the rake face of the tool) [43], with kind permission from Elsevier

Another study by Kaynak et al. [43] mainly addresses the tool wear issue in turning by employing cryogenic, dry, and MQL machining conditions. Two categories of experiments, first one to investigate tool wear at various levels of cutting speed (12.5–25–50–100 m/min) and second one on progressive tool wear at constant cutting speed at 25 m/min were conducted. Figure 2.15 presents the progression of maximum notch wear with cutting speeds under dry, MQL, and cryogenic cooling conditions. Increased cutting speed, beyond 25 m/min, leads to extremely high tool wear and reduced tool life particularly in dry and MQL conditions, whereas cryogenic cooling has significantly reduced the cutting temperature and thereby tool wear and failure, and hence ensured the absence of adverse thermal effects. It is also reported that the cryogenic cooling provides the longest steady-state maximum flank wear region at the tool nose due to the protection of cutting tool against rapid tool wear.

The results show that cryogenic cooling is the best option among the three conditions adopted to improve the machinability of NiTi SMAs. Surprisingly, the results obtained at MQL conditions are ineffective and suggest the use of either cryogenic or MQL coupled cryogenic conditions to conventionally machine the SMAs.

Some other important literature by Kaynak et al. [44, 45] reported on the enhancement in surface integrity of NiTi under the influence of various cooling and lubrication techniques. Turning of NiTi alloys was performed at 0.5 mm/rev feed rate, 0.5 mm depth of cut, and two values of cutting speed, i.e., 12.5 and 100 m/min [44]. It was investigated that at high speed, the surface integrity of cryogenically machined samples are much better than the dry samples (see Fig. 2.16). The reduced tool wear and

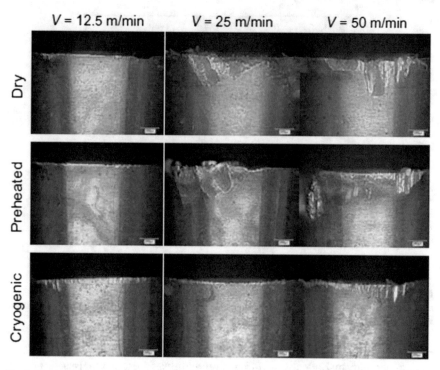

Fig. 2.13 Tool wear patterns at varying cutting speeds and cooling/preheated conditions ($f = 0.1$ mm/rev; $d = 0.5$ mm) [42], with kind permission from Elsevier

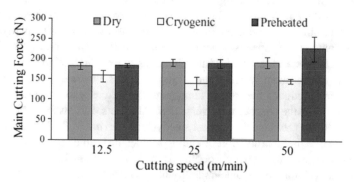

Fig. 2.14 Variation of main cutting force with cutting speeds and cooled/preheated conditions ($f = 0.1$ mm/rev; $d = 0.5$ mm) [42], with kind permission from Elsevier

thermal distortion were recognized to be the main reason behind the generation of smoother surface in cryogenic machining at high cutting speed.

Due to the generation of smoothness and reduced peaks and valleys on the machined surface by cryogenic machining at high cutting speed, surface roughness (0.4 µm) decreased significantly than that of dry machining (1.4 µm). Based

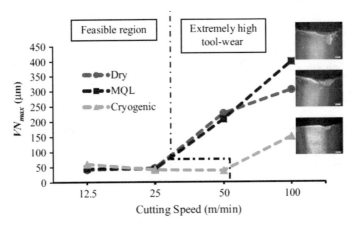

Fig. 2.15 Notch wear progression under different cooling conditions [43], with kind permission from Elsevier

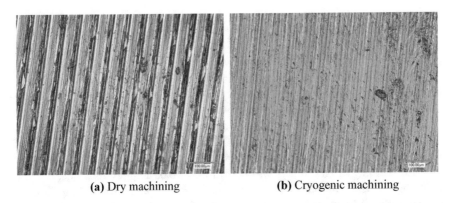

(a) Dry machining   (b) Cryogenic machining

Fig. 2.16 Surface topography of machined NiTi alloys at 100 m/min [44], with kind permission from Elsevier

upon the differential scanning calorimetry (DSC) measurement and analysis of the machined samples, Kaynak et al. [45] observed that martensite to austenite transformation temperatures are higher and transformation peak is broader (see Fig. 2.17) in cryogenically machined sample than the dry machined sample; consequently, residual stress and dislocation density on the surface and subsurface of cryogenically machined sample are expected to be much greater than dry machined sample. This proves the pronounced effect of cryogenic machining on surface integrity characteristics of NiTi alloys.

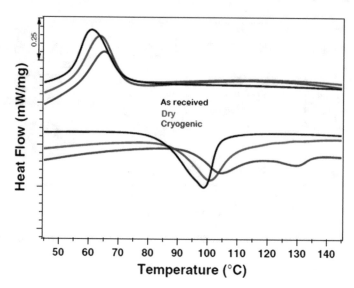

Fig. 2.17 The comparison of the DSC responses of as-received; dry machined and cryogenically machined specimens [44], with kind permission from Elsevier

Results show that cryogenic cooling is the best option among the three conditions adopted to improve the machinability of NiTi SMAs. Surprisingly, the results obtained at MQL conditions are ineffective and suggest the use of either cryogenic or MQL coupled cryogenic conditions to conventionally machine the SMAs.

The attempts made to enhance the surface integrity parameters and the effect of various cooling and lubrication techniques on them are reported by Kaynak et al. in [44, 45]. Turning of NiTi alloys was performed at 0.05 mm/rev feed rate, 0.5 mm depth of cut, and two values of cutting speed, i.e., 12.5 and 100 m/min [44]. As shown in Fig. 2.17, there is not much difference in the surface topographies of the samples machined using dry and cryogenic conditions at low cutting speed. Whereas at high-speed, cryogenically machined samples are much better than the dry samples Fig. 2.18. The smoother surface in cryogenic machining at high cutting speed can be attributed to reduced tool wear and reduced thermal distortion.

Due to the generation of smoother surface and reduced peaks and valleys on the machined surface by cryogenic machining at high cutting speed, surface roughness (0.4 μm) decreased significantly than that of dry machining (1.4 μm). Based upon the DSC measurement and analysis of the machined samples, Kaynak et al. [45] observed that martensite to austenite transformation temperatures is higher and transformation peak is broader (see Fig. 2.19) in cryogenically machined sample than the dry machined sample; consequently, residual stress and dislocation density on the surface and subsurface of cryogenically machined sample are expected to be much greater than dry machined sample. This proves the pronounced effect of cryogenic machining on surface integrity characteristics of NiTi alloys.

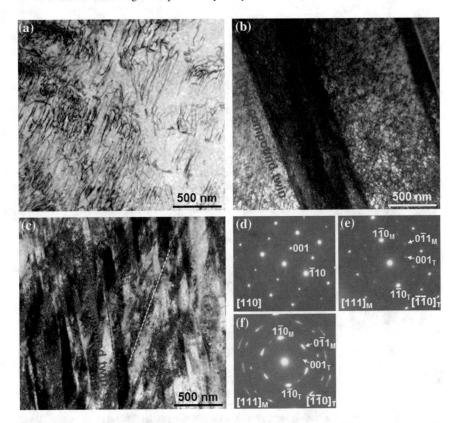

Fig. 2.18 Bright-field transmission electron micrographs of **a** as-received, **b** dry machined and **c** cryogenically machined specimens. **d, e** and **f** are selected-area diffraction patterns obtained from **a, b** and **c**, respectively [45], with kind permission from Elsevier

Some experiments were conducted at 6.25 m/min [45]. The transmission electron micrographs representing the microstructure of as-received material, dry machined specimen, and cryogenically machined sample are shown in Fig. 2.18 a–c that depicts the bright field images of dry and cryogenic samples, respectively. The density of twin bands is higher in cryogenically machined specimen as compared with the dry one. The cryogenic sample also had a greater dislocation density. Much high subsurface hardness distribution is justified as a reason for this phenomenon.

Their complete investigation reveals that cryogenic machining substantially alters the surface integrity of NiTi alloy along with a significant reduction in cutting forces and tool wear. The outcomes of the study conducted by Kaynak et al. encourage further exploration of cryogenic coupled machining for other types of SMAs and machining operations.

Experimental work conducted by Zailani and Mativenga [46] employed chilled air cooling, MQL, and mixture of both with an aim to facilitate the micro-milling of NiTi shape memory alloys. Their major objective was to keep the SMA at low

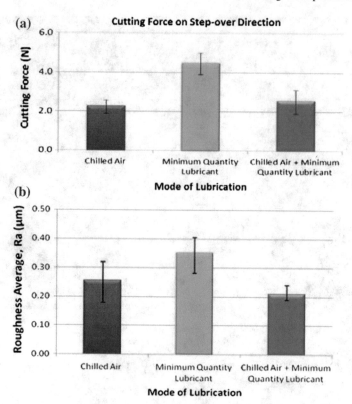

Fig. 2.19 Comparison of **a** cutting force on step-over direction under different cutting conditions, and **b** roughness average under different cutting conditions, in the work conducted by Zailani and Mativenga [46], with kind permission from Elsevier

temperature so that it does not enter the austenite start phase from martensitic phase during machining. Flat end-mill cutters were used for micro-milling of NiTi under the influence of aforementioned cooling and/or lubrication strategies. Machining with chilled air resulted in substantial reduction in grain size and homogeneous microstructure. Moreover, the use of chilled air alone and chilled air concurrently with MQL significantly reduced cutting (feed) forces and improved surface finish (see Fig. 2.19a–b).

Burrs generated under simultaneous application of chilled air and minimum quantity lubricant was smaller and more uniformed in shape. Their concurrent application completely eliminated tool chipping and resulted in the least tool wear (see Fig. 2.20 and 2.21). Overall, it was concluded that the simultaneous use of chilled air and MQL has significant potential and is, therefore, essential to attain most of the machinability indicators in case of SMAs.

In a most recent research, MQL (with Johnson Baby oil) assisted high-speed micro-milling successfully improved the machinability of NiTi alloy in the form

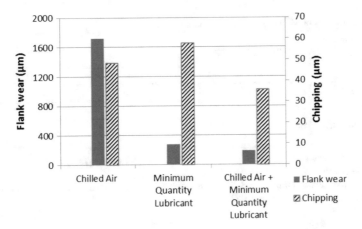

Fig. 2.20 Flank wear and chipping under different cutting conditions, Zailani and Mativenga [46], with kind permission from Elsevier

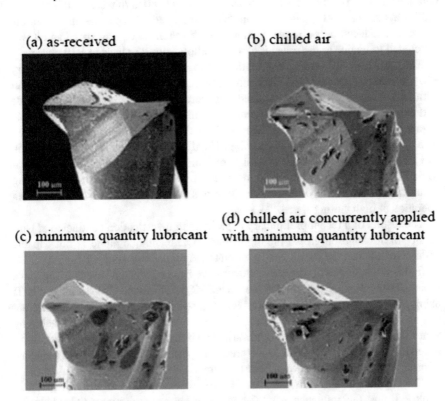

Fig. 2.21 Micro tool **a** as-received; and after machining with **b** chilled air; **c** MQL; **d** chilled air concurrently applied with minimum quantity lubricant [46], with kind permission from Elsevier

Table 2.2 Details of MQL and machining parameters

Variable parameters

Parameter	Unit	Level		
		−1	0	1
Flow rate 'F'	ml/hr	50	70	90
Air pressure 'P'	Bar	20	30	40
Nozzle distance 'D'	mm	4	5	6

Fixed parameters *Cutting speed* 110 m/min; *Feed rate* 0.2 mm/rev; *Depth of cut* 1 mm, Tool material—Carbide

Lubricant properties *Pour point* 8 °C; *Flash point* >290 °C; *Kinematic viscosity* 39.11 mm^2/s at 40 °C; *Density* 0.9199 g/cm^3 at 20 °C

of reduction in machining forces and burr formation during at optimum machining parameters [47]. The air pressure at 4 bar and flow rate 80 ml/h were used for micro-milling at various levels (4.7–30 m/min) of cutting speed. This resulted in low cutting forces and reduced burr size was achieved at 15 m/min of cutting speed and it was concluded to be due to the transition of NiTi alloy from B2 phase to B19 phase.

The assistance of ultrasonic vibration during turning of nitinol shape memory alloys has been explored [48]. Ultrasonic generator (1.2 kW power) and transducer with frequency of 21.5 kHz were used to generate vibrations of 10 μm amplitude at the tip of the turning tool. After employing ultrasonic vibrations, a considerable improvement (up to 70%) in surface finish of the NiTi workpieces was observed than that of the plain conventional turning. The dynamic stiffness of workpiece, tool, and machine set as well as the omission of build-up edge in the presence of ultrasonic vibration were justified as the main reason for the enhancement in machining accuracy. The assistance of ultrasonic vibrations also ensured the successful turning operation at high cutting speed.

The results of a recent preliminary experimental investigation conducted at University of Johannesburg also highlight that MQL technique alone is not capable enough to enhance the machinability of NiTi shape memory alloys.

In this work, a total of nine experiments have been conducted based on Taguchi's robust design of experiment technique during MQL influenced CNC turning of NiTi using plain carbide cutting tool inserts. Micro-droplets of Green lubricant at different settings of flow rate, pressure, and nozzle distance have been supplied in machining zone. Table 2.2 presents the fixed and variable parameters details during MQL influenced turning of NiTi. Figure 2.22a–b present experimental setup used and sequence of tasks performed in the current investigation.

The experimental results, i.e., average roughness and flank wear as given in Table 2.3 show that influence of MQL conditions has not been succeeded to keep the tool flank wear within ISO limit (0.6 mm highest flank wear) and corresponding to all nine experimental combinations its values are much higher than the 0.6 mm, i.e., prescribed ISO limits [49]. The surface finish obtained at NiTi sample is acceptable

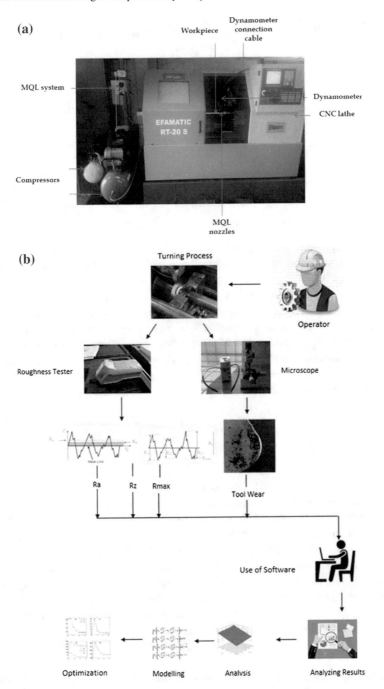

Fig. 2.22 a Experimental setup **b** Sequence of tasks performed during MQL-assisted machining of NiTi shape memory alloy

Table 2.3 Experimental results

Expt. no	F (ml/hr)	P(bar)	D (mm)	R_a (μm)	T_w (mm)
1	50	4	20	1.45	0.76
2	50	5	30	1.12	1.41
3	50	6	40	3.84	0.82
4	70	6	20	1.09	0.80
5	70	4	30	1.17	0.76
6	70	5	40	1.29	1.63
7	90	5	20	1.96	1.39
8	90	6	30	1.03	0.98
9	90	4	40	1.57	1.28

and can be improved further by optimization. Figures 2.23 and 2.24 present the effect of MQL parameters on surface roughness and tool wear. It is seen that the variation of responses with MQL parameters are different, i.e., trends are conflicting in nature or in other words a trade-off exists. For that, multi-objective optimization has been done using desirability analysis technique as an attempt to further improve the surface finish and tool wear and to get single set of parameters for the best machinability. Multi-objective optimization has identified flow rate- 70 ml/hr, 6 bar and 30 mm as optimal parameter combination which produced 1.39 μm average roughness and 1.6 mm tool wear. Even after optimization, the flank wear value is still 1.6 mm that is extremely high and above the ISO prescribed limit for tool failure.

2.4 Summary

The attempts on conventional machining of shape memory alloys (SMAs) reveals that the machining of SMAs without adopting specific advanced machining strategies is extremely difficult. Optimization of machining parameters, utilizing coated tools, and implementing advanced cooling and lubrication strategies, etc., are therefore essential for high surface integrity and reduced tool wear ensuring the manufacture of quality SMAs parts at minimum expenditure and with low environmental footprints. Specifically, the use of advanced cooling and lubrication strategies have significant potential and require sincere future efforts in order to overcome the difficulties of machining various SMAs. Furthermore, it is concluded that the advanced machining and enhancement in the machinability of SMAs is still in its exploration stage, and much scope exists to explore these techniques further to produce high-quality parts of SMAs.

Following may be some possible avenues for future research on machining of SMAs:

Fig. 2.23 Effect of MQL
parameters on average
roughness of NiTi material

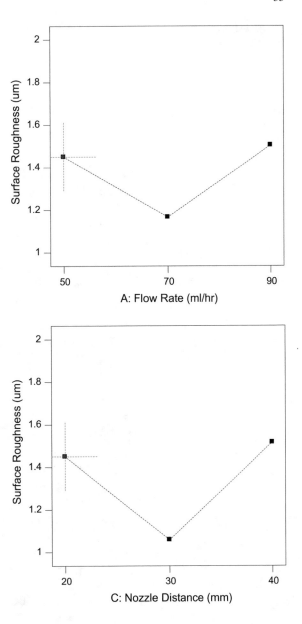

- Conventional machining of shape memory alloys other than nitinol (NiTi) alloy
  such as coper based alloys and Kovar, etc.
- Research focus on conventional processes other than turning and milling, i.e.,
  drilling and grinding, etc.
- More detailed research on hybrid cooling–lubrication, i.e., MQL integrated cryogenic machining.

Fig. 2.24 Effect of MQL
parameters on tool flank
wear

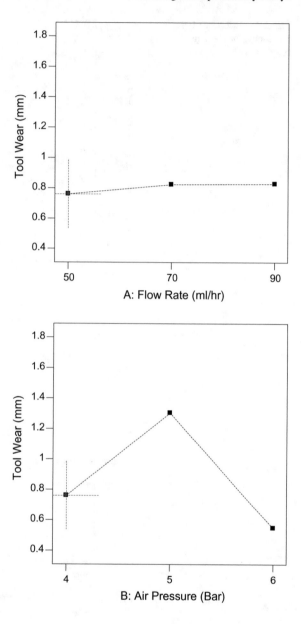

- Extensive study on heat- and vibration-assisted machining.
- Hybrid machining, i.e., combination of advanced cooling or lubrication and heat-
  and/or vibration-assisted machining, etc.
- Advanced machining of shape memory alloys other than NiTi.

References

1. Y. Guo, A. Klink, C. Fu, J. Snyder, Machinability and surface integrity of Nitinol shape memory alloy. CIRP Ann. **62**, 83–86 (2013)
2. M.R. Hassan, M. Mehrpouya, S. Dawood, Review of the machining difficulties of nickel-titanium based shape memory alloys. Appl. Mech. Mater. **564**, 533–537 (2014)
3. K. Weinert, V. Petzoldt, Machining of NiTi based shape memory alloys. Mater. Sci. Eng. A **378**, 180–184 (2004)
4. K. Gupta, R.F. Laubscher, Sustainable machining of titanium alloys: a critical review. Proc. Inst. Mech Eng Part B J. Eng. Manuf. **231**(14), 2543–2560 (2017)
5. E.O. Ezugwu, Key improvements in the machining of difficult-to-cut aerospace superalloys. Int J Mach Tool Manu **45**(12–13), 1353–1367 (2005)
6. K. Gupta, N.K. Jain, R.F. Laubscher Jain, Spark-erosion machining of miniature gears: a critical review. Int. J. Adv. Manuf. Technol. **80**(9–12), 1863–1877 (2015)
7. K.P. Rajurkar, M.M. Sundaram, A.P. Malshe, Review of electrochemical and electro discharge machining. Procedia CIRP **6**, 13–26 (2013)
8. K. Gupta, N.K Jain, Chapter-2 Overview of wire spark erosion machining. in *Near Net Shae Manufacturing of Miniature Spur Gears* (Springer, 2016)
9. H.C. Lin, K.M. Lin, I.S. Cheng, The electro-discharge machining characteristics of TiNi shape memory alloys. J. Mater. Sci. **36**(2), 399–404 (2001)
10. S.L. Chen, S.F. Hsieh, H.C. Lin, M.H. Lin, J.S. Huang, Electrical discharge machining of a NiAlFe ternary shape memory alloy. J. Alloy. Compd. **464**, 446–451 (2008)
11. M.H. Abidi et al., An investigation of the micro-electrical discharge machining of nickel-titanium shape memory alloy using grey relations coupled with principal component analysis. Metals **7**, 486 (2017). https://doi.org/10.3390/met7110486
12. S. Daneshmand, E.F. Kahrizi, A.A.L. Neyestanak, M. Mortazavi, Experimental investigations into electro discharge machining of NiTi shape memory alloys using rotational tool. Int. J. Electrochem. Sci. **8**, 7484–7497 (2013)
13. S. Daneshmand, E.F. Kahrizi, A.A.L. Neyestanak, V. Monfared, Optimization of electrical discharge machining parameters for NiTi shape memory alloy by using the Taguchi method. J. Marine Sci. Technol.-Taiwan **4**, 506–512 (2014)
14. S. Daneshmand, V.M.A. Akbar, Lotfi Neyestanak, Effect of tool rotational and Al_2O_3 powder in electro discharge machining characteristics of NiTi-60 shape memory alloy. Silicon **9**(2), 273–283 (2017)
15. M. Mallaiah, S. Narendranath, S. Basavarajappa, V.N. Gaitonde, Influence of process parameters on material removal rate and surface roughness in WED-machining of Ti50Ni40Cu10 shape memory alloy. Int. J. Mach. Mach. Mater. **18**(1/2), 26–53 (2016)
16. M. Manjaiah, R.F. Laubscher, S. Narendranath, S. Basavarajappa, V.N. Gaitonde, Evaluation of wire electro discharge machining characteristics of Ti 50 Ni 50-xCux shape memory alloys. J. Mater. Res. **31**(12), 1801–1808 (2016)
17. M. Manjaiah, S. Narendranath, S. Basavarajappa, V.N. Gaitonde, Investigation on material removal rate, surface and subsurface characteristics in wire electro discharge machining of Ti50Ni50-xCux shape memory alloy. Proc. Inst. Mech. Eng. Part L J. Mate. Design Appl. **232**(2), 164–177 (2018)
18. M. Mallaiah, S. Narendranath, J. Akbari, Optimization of wire electro discharge machining parameters to achieve better MRR and surface finish. Procedia Mater. Sci. **5**, 2635–2644 (2014)
19. M. Mallaiah, S. Narendranath, S. Basavarajappa, V.N. Gaitonde, Effect of electrode material in wire electro discharge machining characteristics of Ti50Ni50-xCux shape memory alloy. Precision Eng. **41**, 68–77 (2015)
20. J. Hecht, Short history of laser development. Opt. Eng. **49**(9), 1002 (2010)
21. K.C. Yung, H.H. Zhu, T.M. Yue, Theoretical and experimental study on the kerf profile of the laser micro-cutting NiTi shape memory alloy using 355 nm Nd: YAG. Smart Mater. Struct. **14**, 337–342 (2005)

22. R. Pfeifer, D. Herzog, M. Hustedt, S. Barcikowski, Pulsed Nd: YAG laser cutting of NiTi shape memory alloys-influence of process parameters. J. Mater. Process. Tech. **210**(14), 1918–1925 (2010)

23. C.H. Hung, F.Y. Chang, T.L. Chang, Y.T. Chang, K.W. Huang, P.C. Liang, Micromachining NiTi tubes for use in medical devices by using a femtosecond laser. Opt. Lasers Eng. **66**, 34–40 (2015)

24. C. Li, S. Nikum, F. Wong, An optimal process of femtosecond laser cutting of NiTi shape memory alloy for fabrication of miniature devices. Opt. Lasers Eng. **44**, 1078–1087 (2006)

25. C.A. Biffi, A. Tuissi, Fiber laser drilling of $Ni_{46}Mn_{27}Ga_{27}$ ferromagnetic shape memory alloy, *Optics Laser Technol.* **63**, 1–7 (November 2014)

26. T. Phokane, K. Gupta, M.K. Gupta, Investigations on surface roughness and tribology of miniature brass gears manufactured by abrasive water jet machining. Proc. Inst. Mech. Eng. Part C J. Mech. Eng. Sci. 0954406217747913

27. M. Hashish, Abrasive waterjet cutting of microelectronic components. in *Proceedings of the 2005 WJTA American Waterjet Conference*, vol. 26 (Houston, Texas, USA, August 2005)

28. M. Frotscher, H. Gugel, K. Neuking, W. Theisen, G. Eggeler, F. Kahleyß, D. Biermann, Machining of stent-like geometries in thin NiTi sheets using water jet cutting. ed. by J. Gilbert. *Medical Device Materials V, Proceedings from the Materials & Processes for Medical Devices Conference*, (USA, 2009), pp. 10.8.–12.8, 201–206

29. M.C. Kong, D. Axinte, W. Voice, Challenges in using waterjet machining of NiTi shape memory alloys: An analysis of controlled-depth milling. J. Mater. Process Tech. **211**, 959–971 (2011)

30. M.C. Kong, D. Srinivasu, D. Axinte, W. Voice, J. McGourlay, B. Hon, On geometrical accuracy and integrity of surfaces in multi-mode abrasive waterjet machining of NiTi shape memory alloys. CIRP Ann. **62**(1), 555–558 (2013)

31. H.C. Lin, K.M. Lin, Y.C. Chen, A study on the machining characteristics of TiNi shape memory alloys. J. Mater Process. Tech. **105**(3), 327–332 (2000)

32. K. Weinert, V. Petzoldt, D. Kötter, Turning and drilling of NiTi shape memory alloys. CIRP Ann. **53**(1), 65–68 (2004)

33. K. Weinert, V. Petzoldt, Deep hole drilling of NiTi shape memory alloys. in *SMST: Proceedings of the International Conference on Shape Memory and Superelastic Technologies* (2006), pp. 259–264

34. K. Weinert, V. Petzoldt, Machining NiTi micro-parts by micro-milling. Mater. Sci. Eng., A **481**, 672–675 (2008)

35. K. Gupta, R.F. Laubscher, J. Paulo Davim, N.K. Jain, Recent developments in sustainable manufacturing of gears: a review. J. Cleaner Prod. **112**(4), 3320–3330 (2016). (Elsevier)

36. K. Gupta, R.F. Laubscher, MQL assisted machining of grade-4 titanium. in *Proceedings of International Conference on Competitive Manufacturing (COMA)*, (Stellenbosch, (South Africa), 27–29, January 2015), pp 211–217

37. D. Biermann, F. Kahleyss, T. Surmann, Micromilling of NiTi shape-memory alloys with ball nose cutters. Mater. Manuf. Process **24**(12), 1266–1273 (2009)

38. D. Biermann, F. Kahleyss, E. Krebs, T. Upmeier, A study on micro-machining technology for the machining of NiTi: five-axis micro-milling and micro deep-hole drilling. J. Mater. Eng. Perform. **20**(4–5), 745–751 (2011)

39. R. Piquard, A. D'Acunto, D. Dudzinski, Study of burr formation and phase transformation during micro-milling of NiTi alloys. in *11th International Conference on High Speed Machining*, (Czech Machine Tool Society, 2014), pp. 1–6

40. R. Piquard, A. D'Acunto, P. Laheurte, D. Dudzinski, Micro-end milling of NiTi biomedical alloys, burr formation and phase transformation. Precis. Eng. **38**(2), 356–364 (2014)

41. I. Shyha, M. Patrick, I. Elgaly, Machinability analysis when drilling Kovar shape memory alloys. Adv. Mater. Sci. Eng. **1**(3–4), 411–422 (2015)

42. Y. Kaynak, R.D. Noebe, H.E. Karaca, I.S. Jawahir, Analysis of tool-wear and cutting force components in dry, preheated, and cryogenic machining of NiTi shape memory alloys. Proc. CIRP **8**, 498–503 (2013)

43. Y. Kaynak, H.E. Karaca, R.D. Noebe, I.S. Jawahir, Tool-wear analysis in cryogenic machining of NiTi shape memory alloys: a comparison of tool-wear performance with dry and MQL machining. Wear **306**, 51–63 (2013)

44. Y. Kaynak, H.E. Karaca, I.S. Jawahir, Surface integrity characteristics of NiTi shape memory alloys resulting from dry and cryogenic machining. Proc. CIRP **13**, 393–398 (2014)

45. Y. Kaynak, H. Tobe, R.D. Noebe, H.E. Karaca, I.S. Jawahir, The effects of machining on the microstructure and transformation behavior of NiTi Alloy. Scripta Mater. **74**, 60–63 (2014)

46. A.Z. Zailania, P.T. Mativenga, Effects of chilled air on machinability of NiTi shape memory alloy. Procedia CIRP **45**, 207–210 (2016)

47. R. Kuppuswamy, A. Yui, High-speed micromachining characteristics for the NiTi shape memory alloys. Int. J. Adv. Manuf. Technol. **93**(1–4), 11–21 (2017)

48. J. Akbari, A.G. Chegini, A.R. Rajabnejad, *Ultrasonic-Assisted Turning of NiTi Shape Memory Alloy, 5th International Conference and Exhibition on Design and Production of MACHINES and DIES/MOLDS 18–21 JUNE 2009* (Pine Bay Hotel—Kusadasi, Aydin, Turkey, 2009)

49. ISO 3685, *Tool Life Testing with Single-point Turning Tools* (Int. Organization Stand., Geneva, 1977)

Chapter 3
Welding and Joining of Shape Memory Alloys

3.1 Introduction

Welding and joining is a vital classification of manufacturing processes, wherein two or more different parts are joined together by the cause of fusion or mechanical pressure or mechanical fasteners or adhesive action or different combinations of these causes [1]. Welding and joining of shape memory alloys are challenging as the effects of shape memory alloys must remain as it is after its welding and joining [2]. In addition to this, the behavior of shape memory alloy is different during the process performance as the welding and joining processes are performed with an application of certain mechanical pressure and large amount of heat, which subsequently leads to the difficulties in obtaining sound joints [2, 3]. However, different welding and joining processes such as tungsten inert gas welding (TIG), plasma welding, laser beam welding (LBW), electron beam welding (EBW), resistance welding, friction stir welding (FSW), friction welding, explosive welding, ultrasonic welding, diffusion bonding, adhesive bonding, brazing and soldering are reported [2, 3] in the different literatures that are utilized to obtain different shape memory alloys (see Fig. 3.1 for classification). These processes are further discussed in detail based on three different categories such as joining, fusion and beam welding, and solid state welding.

3.2 Fusion and Beam Welding Processes

Fusion welding is a category of welding, in which the base materials are joined by means of melting of material caused by intense localized heat. This supply of heat may be arc or beam or resistance heat that determines its subclassification such as fusion arc welding processes, resistance welding, and fusion beam welding processes. TIG and plasma welding processes that fall under the category of fusion arc welding are reported as applied processes to obtain joints of shape memory alloys [2, 3]. In case

K. Mehta and K. Gupta, *Fabrication and Processing of Shape Memory Alloys*,
Manufacturing and Surface Engineering, https://doi.org/10.1007/978-3-319-99307-2_3

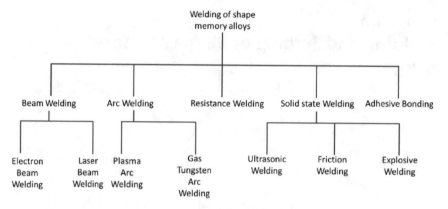

Fig. 3.1 Classification of welding processes applied to weld shape memory alloys

of fusion beam welding, electron beam welding and laser beam welding processes are applied to achieve weld of shape memory alloys [2, 3]. Process-wise explanation is presented below.

3.2.1 TIG Welding

TIG welding uses non-consumable tungsten electrode from where an electric current is passed, in order to generate the arcing effect. Due to arc, the base material gets melted and subsequently solidifies as the time progress that in turn leads to the joining. TIG welding process has advantages such as easy to operate process, low heat input, simple welding source, and applicable to weld thin materials, which attracts it to the range of applications [1–3]. Joining of shape memory alloys with TIG welding is challenging as it involves melting of material [2, 3]. The challenges of embrittlement, formation of intermetallic compounds (IMCs), solidification cracking and retention of shape memory effects are reported with TIG welding of shape memory alloys. It is also found that post welding treatments are required to overcome these challenges up to some extent. TIG welding has advantages of any coupling applicability, easy welding procedure, and arc stability at small current, easy to apply on thin workpiece, which makes it suitable to employ for welding of shape memory alloys [2, 3]. History on welding of shape memory alloys tells that TIG welding process is the first process that applied on NiTi alloy in 1961 [2]. In this study, they reported that interstitial phases based on H, N, and O can be observed because of insufficient protection of molten pool. Ikai et al. [4] have communicated that the mechanical properties of NiTi welds are noted as decreased than base material. Two different configurations such as NiTi wire of 0.75 mm diameter and 0.2 mm thick sheet are welded with the help of TIG welding. The favorable welding conditions of TIG welding for NiTi joints of 0.75 mm diameter wire are reported as TIG current of 10 A, direct current and plus polarity, 1.47 N of wire push force, 1 mm of wire push length, and 1.6 mm of electrode diameter. Besides, welding conditions of direct current and plus polarity, 4 A of welding current and 1 mm of electrode diameter are reported as favorable conditions for 0.2 mm thick NiTi sheets. The heat-affected zone of TIG-welded specimens is reported as easily deformable compared to the base material and weld zone. They reported that strain recovery is highly poor during cyclic tests of 50 cycles at 4% strain. The welded wires are observed to perform full cyclic test of 50 cycles, whereas welded sheets got ruptured after 39 cycles.

NiTi is prone to react with oxygen, hydrogen, and nitrogen at high temperature, therefore, TIG welding may lead to forming super brittle weld zone. Recently, Oliveira et al. [5] implemented a TIG welding with fixing device of inert gas flowing system at root and face of the welds on 1.5 mm thick NiTi plates for butt joint configuration, which has minimized the effect of oxidation due to inert gas flowing protection. This method, in turn, showed 20% rupture of total strain and superelastic behavior with 30 MPa strength below the base material. Here, cyclic superelasticity of stain up to 12% and 600 cycles are performed. The shape memory effect is reported as retained even after the welding. Besides, the transformation temperature is reported as increased for weld compare to base material due to presence of Ti oxide, despite the prevention by argon fixation is implemented. Therefore, it can be stated that proper use of shieling gas and prevention backing gas is the key parameter for TIG welding of shape memory alloys.

Dissimilar welding of shape memory alloys with another metal/alloy is performed by TIG welding process such as NiTi to stainless steel (SS) 304 and NiTi to SS 316L. NiTi to SS 304 is carried out using Ni insert for the tube to tube joint configuration [6, 7]. The tube was having thickness of 1.9 mm and an outer diameter of 9.3 mm. Ni insert acted as filler material to weld NiTi to SS 306 through TIG welding [6]. The mechanical properties such as hardness of 817 HV and ultimate torsional strength of 415 MPa reported as satisfactory along with narrow width of heat-affected zone (HAZ), considering the application of NiTi-SS 304 joints for the actuators. Formation of IMCs such as Ni_3Ti_2 and Ni_4Ti_3 are reported inside the weld zone of dissimilar NiTi-SS 304 TIG welds. Therefore, the weld zone of NiTi-SS 304 joints causes higher hardness. TIG welding of NiTi to SS 304 results into full penetration in a single pass besides of larger HAZ due to large bead width of weld caused through higher energy relative to laser welding technique [6].

TIG welding is interestingly utilized to join NiTi of Ti reach wires for the medical occluders application, which is further connected with stainless steel pipe through laser spot welding as shown in Fig. 3.2 [7]. TiC is reported at the welds of wire while IMCs such as $Ni_3Ti + (Fe, Ni)$ Ti observed around the fusion zone of NiTi-SS 304 laser spot weld. Furthermore, in applications, the overlay of NiTi on carbon steel is carried out with the TIG welding. Benefits of increase in hardness are reported due to formation of oxides and IMCs with superelastic maintained effect. The erosion effect is drastically improved over a wide range of impact angle through this overlay of NiTi on carbon steel by TIG welding [7].

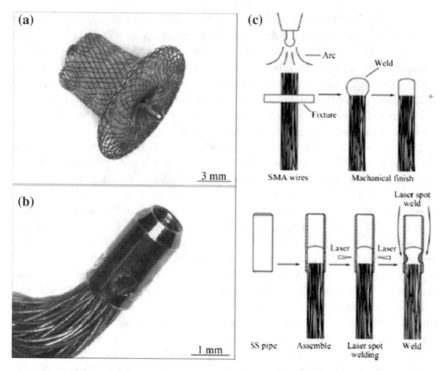

Fig. 3.2 Medical occluders **a** TIG-welded NiTi wires, **b** laser spot welded assembly of pipe to wires and **c** manufacturing procedure TIG of wires and laser spot welding of wires to pipe [7], with kind permission from Elsevier

3.2.2 Plasma Welding

Plasma welding uses non-consumable electrode of tungsten similar to TIG and the process principle is also matching with TIG process. The difference is extra shielding which leads to separate the plasma arc in addition to shielding gas [8]. Plasma welding is utilized to weld NiTi shape memory alloy, NiTi-stainless steel, and Niti-Hastalloy by van der Eijk et al. [9]. Plasma welding with filler of powder is applied via robotic welding station. Argon gas is utilized for plasma gas, outer shielding, and backing. Investigations summarized that transformation temperatures of NiTi–NiTi weld are retained while mechanical properties are deteriorated. Plasma welding of NiTi–NiTi welds has not affected the ratio of Ni and Ti materials. On the other hand, fusion line of NiTi-Hast alloy and NiTi-stainless steel has got brittle phases and defects. NiTi has absorbed different elements from Hast alloy and stainless steel, which is not desirable. In short, plasma welding is not strongly recommended for shape memory alloys. NiTi-Hast alloy weld resulted in six different phases that are identified by scanning electron microscopy (SEM) image and EDX analysis. Julien et al. [10] have developed a plasma spray technique to coat a Nitinol on metallic substrate.

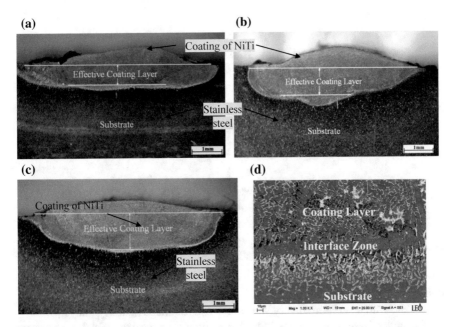

Fig. 3.3 NiTi coating on Stainless steel substrate by plasma arc transferred method **a** 80 A, **b** 90 A, **c** 100 A and **d** SEM image [11], with kind permission from Elsevier

Nitinol is used in a powder form that is mixed with hydrogen and argon gases and converted to partially molten state by ionizing and heating. They achieved sound diffusion bonding between nitinol IMC and metallic substrate via high velocity and impact of partially melted powder of nitinol. Ozel et al. [11] have done experiments on NiTi coating on stainless steel with the help of plasma transferred arc process. They reported coating layer of 1, 1.2, and 1.4 mm as effective at 80, 90, and 100 A currents, respectively without any defects as shown in Fig. 3.3. The presence of defects is often reported with coating by TIG applications that problem is exceeded by plasma arc transferred method.

3.2.3 Laser Beam Welding

Laser beam welding uses a beam of light through optical amplification based on stimulated emission of radiation to melt the base material and that leads to the bonding after its solidification [8]. Among all the available welding techniques, laser welding is reported as most employed technique for the shape memory alloys considering its extraordinary benefits such as better process control, low heat input, high density, ability to reproduce, low HAZ, and monochromatic nature [1–3]. It is also well

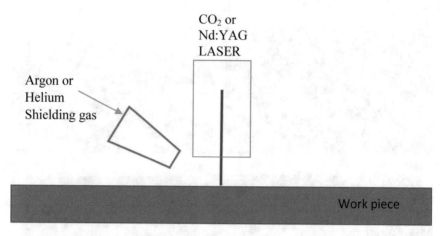

Fig. 3.4 Laser welding setup with shielding gas for shaper memory alloys

documented that the laser beam welds are having narrower welding zone than the arc welds, which is reported very well in Fe–Mn–Si shape memory alloys [2].

The type of laser such as CO_2, Nd:YAG, or fiber laser used to weld shape memory alloys significantly affects the weld properties [2, 3, 12–31]. However, the Nd:YAG and fiber laser are considered as priority one relative to the CO_2 lasers [2]. The reason for this is low wavelength, which leads to increase of absorptivity and consequently form narrow welds. The laser welding process parameters of laser focus size and position, wavelength, power and mode, welding speed, type of shielding gas and flow rate are reported as most influencing parameters in case of shape memory alloys welding [2]. It is suggested from the literatures that the pulse waveform of laser is preferred for welding wire of thin foils while continuous waveform of laser is considered for high thickness shape memory alloys [2, 3, 12–31]. Laser welding of NiTi material can be effectively done when shielding gas is supplied [2]. The use of shielding gas in laser welding of shape memory alloys is illustrated in Fig. 3.4. It is reported that shielding gases such as argon and helium are used for NiTi shape memory material, which minimizes oxidation and contamination during welding. Argon is also used as shielding gas for other shape memory alloys such as Cu–Al–Mn, in the case of laser welding [13].

Laser-welded shape memory alloy joints are observed as better in terms of preserving shape memory effect [12]. It is reported that at least 90% of the shape memory effect is preserved by recovering deformation. Several articles report that full recovery on shape memory effect of imposed strain can be achieved on welds obtained by laser welding. The recovery of shape memory effect is dependent on stain rate. It is reported that the irrecoverable strain is consisted after martensitic to austenitic transformation with the applied strain. However, this irrecoverable strain is reported as small as below 0.1 and 0.3% for stain of 4 and 6%, respectively, relative to the base material of NiTi material. It is well experimented by Falvo et al. [14] that, the

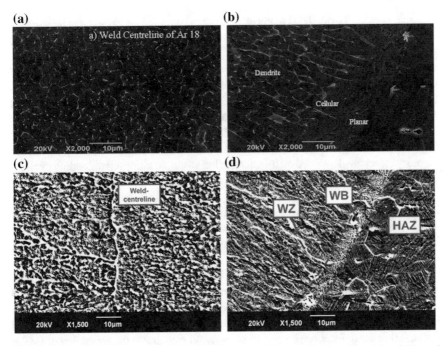

Fig. 3.5 Microstructures of laser-welded shape memory alloys, weld center line: (**a**) and (**c**), and weld interface boundary: (**b**) and (**d**) [15, 23], with kind permission from Elsevier

laser-welded Ti-rich NiTi has 4% strain recovery for the 5% imposed strain. Besides, they reported that 5.2% of recovery of strain is obtained for the 7.3% of imposed strain in the case of the base material.

Laser-welded shape memory alloys consist of interesting microstructures inside the weld zone and HAZ. The weld zone microstructures are reported with dendritic type, planner, cellular or combination of any/all of these microstructures [15, 16, 23]. Dendritic type structure is reported due to rapid solidification theory. The changes of solidification structures such as planner to cellular, and cellular to dendritic occur because of increase in constitutional supercooling of shape memory alloy. Therefore, the possibility of existing one microstructure among any one or two aforementioned structures or all the structures is maximum [15, 16, 23]. One of the examples of laser-welded NiTi alloy is shown in Fig. 3.5 for its different reported microstructures. In most of the cases, the HAZ of shape memory alloys is consisted recrystallized equiaxed grains microstructure with an increased grain size because of heat conduction leads gradient of recrystallization.

Different reports have reported that, at the room temperature, the microstructures of austenite and martensitic are coexisted inside the weld and HAZ of laser-welded NiTi, whereas fully austenitic structure is presented at the base material [21–31]. The synchrotron X-ray radiation and conventional CuKσ radiation technologies are utilized to identify these structural changes [15, 16].

Laser-welded Ti-rich NiTi material is reported with the decreased tensile strength of welds due to the presence of microcracks at the weld zone and grain boundary precipitation of Ti_2Ni at HAZ [15, 16, 31]. Precipitation leads to the embrittlement, which causes lower ductility and lower strength. The lower ductility is reported with laser pulse welded NiTi material that is because of dislocations in the weld zone [2]. Besides, in the case of Cu–Al–Mn shape memory alloy, no significant difference of tensile properties is observed relative to base material as the grain size of weld is not much affected. The ductility of laser-welded Cu–Al–Mn material is also reported higher and necking is observed during tensile test [12, 13]. Tensile testing and shape memory effect are depended on transformation strain, loading and unloading cycling, shape memory effect recovery, superelastic recovery, and start and finishing stresses [2].

The application of the laser welding for the dissimilar materials joining (where shape memory alloy is one of the materials) is reported in a large number of articles due to process advantages of the high power density, reduced thermally affected regions, easy control, flexibility, and reproducibility. Dissimilar combinations of NiTi-steel, NiTi-stainless steel, NiTi-titanium, NiTi-Monel, NiTi-Inconel, NiTi-Tantalum, NiTi-Cu, NiTi-CuAlMn, and NiTi–Cobalt chromium are attempted in different literature with laser welding application [2, 24–26, 31]. The differences in thermo-physical properties lead to the formation of IMCs at the weld interface. It is noted that the defects can be placed in the weld area due to the formation of IMCs. Careful control of process parameters can avoid the problem of IMCs formation. Some of the studies reveal that the use of interlayer in the dissimilar laser welding improves the joint properties by reducing formation of IMCs and defects [2, 24–26, 31]. Various interlayers of nickel, cobalt, copper, iron, and niobium are recommended for different combinations such as NiTi-steel, NiTi-stainless steel, and NiTi-Ti. Most of the dissimilar laser-welded joints are reported with higher hardness values at the joint area. Higher tensile properties than the base material are possible to obtain with dissimilar laser welding. Some of the literature of dissimilar laser welding report that the superelastic behavior and the shape memory effect is possible to achieve with dissimilar welds of shape memory alloy with other material. The cyclic behavior can be retained when laser welding is applied with controlled process parameters on dissimilar combinations having one of the base materials of shape memory alloys [2, 24–26, 31].

3.2.4   Electron Beam Welding

Electron beam welding works on the similar principle of laser beam welding that operates with a beam of electrons. Beam of electrons is focused on the workpiece to get it melt, in the case of electron beam welding. The setup of electron beam welding requires vacuum chamber wherein the electrons are accelerated with high voltage and focuses on workpiece by the lenses of electromagnetic fields. Electron beam is even more advantageous than the laser welding in terms of beam control, prevention

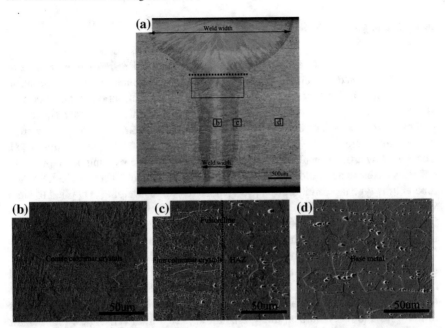

Fig. 3.6 Microstructures of electron beam welded NiTi [32], with kind permission from Elsevier

of oxidation and contamination, small beam diameter of beam and high energy of beam per unit area [8]. However, few research works are reported for the electron beam welding of shape memory alloys as the electron beam is limitedly applied in the industries. This is due to disadvantages of high initial setup cost, high maintenance cost, time required initially for vacuum generation, and operate workpiece in vacuum chamber [2].

Yang et al. [32] performed experiments on electron beam welding of NiTi material for the behavior of welds in terms of microstructure and mechanical properties. They reported 85% of joint efficiency at room temperature, while they obtained 93% joint of efficiency at the temperature of 223 K. NiTi microstructure is reported as columnar crystal type to columnar fine grain type as shown in Fig. 3.6. Columnar grains are observed as perpendicular to the weld center line. Precipitates of Ti_2Ni or Ni_4Ti_3 are noted in the weld zone. Differential scanning calorimetry is showing that only one step reversible martensitic transformation is obtained in weld and base material. No major changes in transformation temperatures are reported in the case of electron beam welding of NiTi. Only, the martensitic start temperature of the weld is reported as increased. The same group of authors Yang et al. [33] reported via different article that post weld annealing on electron beam welded NiTi reduces martensitic start temperature of the weld. Optimum phase transformation can be achieved after post weld annealing with at 1073 K temperature.

3.3 Resistance Welding

Resistance welding is a technique in which the electrical resistance is utilized to provide heat based on Joule's law that leads to material melting and deformation together. The forging force in addition to melting and deformation creates bonding at the interface of the workpiece. The current is passed through copper electrodes in most of the cases. The contributory parameters of resistance welding are current density, time, pressure to forge, and electrical resistance of the base material [8]. There are few articles available which investigate resistance welding of shape memory alloys. Resistance welding is applied on micro welding of NiTi wires [34, 35] and tube to tube type lap configuration [36]. Micro resistance welding is applied to weld wires of 0.4 mm diameter for cross wire joints. The joining mechanism observed in this study is solid state bonding consisted with six stages such as cold collapse, dynamic recrystallization, interfacial melting, squeeze out, excessive flash, and surface melting. The transformation temperature of the weld is reported as modified due to annealing effect on stain hardened base material caused by resistance based heat input. Phase transformation of weld is reported at lower temperature relative to the base material [34]. Microstructure of the weld is consists of columnar grains in the weld and recrystallized fine grains are reported in the HAZ (see Fig. 3.7 for microstructure at interface, weld, and HAZ under different current). Distinct interface of Fig. 3.7b, d is because of improper bonding as the current applied is low, while no such interface is reported at higher currents as shown in Fig. 3.7f, h. Presence of TiC precipitates in weld area is proposed based on microstructural observations [34]. Recrystallized fine grains are reported in large zone with higher current as the HAZ is maximum in case of highest applied current. Welding current is observed for its significant effects on joint properties including strength, Pseudoelasticity, transformation and shape memory effect [34].

Resistance welding is interestingly utilized by Delobelle et al. [36] to form architectural cellular structure of NiTi tube, which is like porous material. Resistance welding is a technique that can be used to fabricate a complex structure of NiTi. Resistance welding affects this joint in terms of transformation behavior, microstructural properties, and mechanical properties. Post welding treatment is suggested to obtain identical transformation behavior of the joint relative to base material. The solution treatment with temperature of 850 °C for 60 min followed by aging at a temperature of 450 °C for 30 min is recommended for resistance welded Ti-50.8 at.% Ni shape memory material. At the same time, application of this heat treatment does not affect hardness, grain size, and modulus of elasticity significantly.

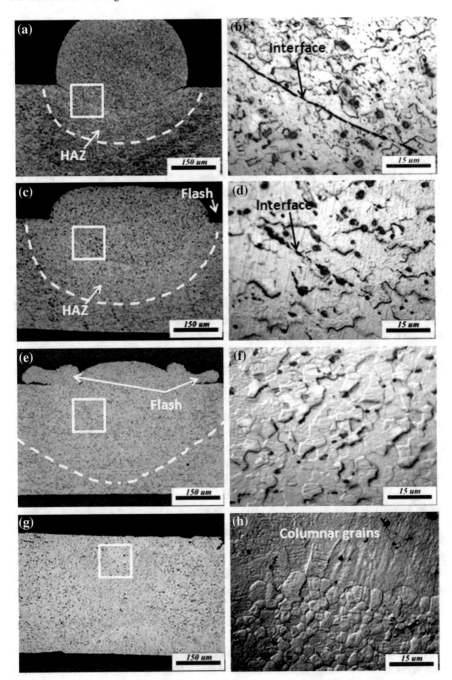

Fig. 3.7 SEM images of resistances spot welded NiTi wires with different currents: **a**, **b** 145 A, **c**, **d** 195 A, **e**, **f** 245 A, **g**, **h** 295 A [34], with kind permission from Springer

3.4 Solid State Welding

Solid state welding processes are those processes, which produce a joint by means of solid-state deformation of base material caused by heat and pressure wherein melting of base material and filler wire addition is not required [1, 8]. Since the material melting is avoided, the solid-state welding processes are applicable where the weld zone is prone to form brittle IMCs [1]. Hence, solid state welding processes are categorized as suitable processes for welding of shape memory alloys [2]. Different solid state welding processes such as friction welding, friction stir welding, ultrasonic welding, and explosive welding are investigated for shape memory alloys [2]. Individual explanations for each of these processes are mentioned as under.

3.4.1 Friction Welding

Friction welding is a type of solid-state welding processes that uses frictional heat and axial pressure to plastically deform the material for a specific time. This plastic deformation leads to the upset of base materials and subsequently causes intermixing to form a solid-state joint. Melting of the base material is not required as the joint formation is due to the solid-state deformation [8]. Friction welding has proved its feasibility to join similar welds of shape memory alloys and dissimilar welds [2]. Despite ability of making shape memory alloys weld, the problems such as substantial change in temperature of phase transformation and lower joint strength are reported. Friction welding is investigated for NiTi material and dissimilar NiTi to stainless steel materials. Friction-welded NiTi joints are consisted of much finer grains than base material due to dynamic recrystallization. It is reported by Shinoda et al. [37, 38] that the phase transformation temperature of a welded zone of NiTi material is not in line with base material. They suggested that proper heat treatment is required to implement after friction welding in order to reduce the variations of phase transformation temperature and shape memory behavior relative to the base material. Heat treatment of NiTi joints after friction welding helps to enhance the tensile strength and shape memory effect relative to base material. It is mentioned that heat treatment before friction welding causes lowering the phase transformation temperatures. It is reported that higher upset pressure is required to obtain successful welds. At least upset pressure of 127 MPa is necessary to obtain 6 mm diameter NiTi friction welds.

Dissimilar friction welding of NiTi to stainless steel is carried out on rod of 2.5 mm diameter with the help of nickel interlayer as shown in Fig. 3.8. Fukumoto et al. [39] have successfully implemented the dissimilar NiTi to stainless steel using Ni interlayer as shown in Fig. 3.9a. It is reported that Ni interlayer has helped to improve the mechanical properties of NiTi to stainless steel dissimilar joints. Maximum tensile strength 512 MPa is obtained using Ni interlayer for dissimilar NiTi to stainless steel friction welds. Formation of IMCs such as Ni_3Ti and $NiTi + Ni_3Ti$ of eutectic layer is

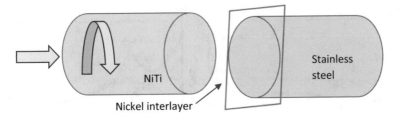

Fig. 3.8 Friction welding concept applied to dissimilar NiTi-stainless steel joints

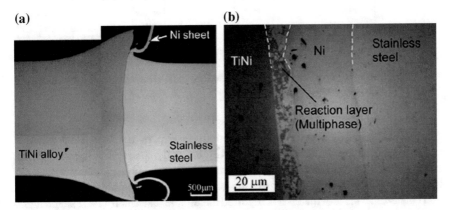

Fig. 3.9 Dissimilar friction-welded NiTi-stainless steel joint, **a** cross-section image and **b** microstructure image with Ni interface [39], with kind permission from Taylor and Francis

reported as reaction layer between NiTi and stainless interface (see Fig. 3.9b). These phases are formed because of inter-diffusion and mechanical mixing phenomenon. Ni interlayer has prevented the formation of Fe_2Ti phase of IMCs. Fe_2Ti is brittle in nature and hence prevention of this phase by applying Ni interlayer has improved the brittleness of the joint. On the other hand, the IMC phase of Fe_2Ti is found in case of friction welding of NiTi to stainless steel material without using Ni interface. Formation of IMCs is depended on process parameters of friction welding. Rotational speed governs eutectic reaction and IMCs formation significantly. As rotational speed increases, the eutectic reaction, and layer of IMCs generated increases.

3.4.2 Friction Stir Welding

Friction stir welding (FSW) is an advanced friction welding type process wherein the frictional heat is generated with the help of an external non-consumable tool that leads to the plastic deformation of the base material [40, 41]. Rotation and transverse moment of the FSW tool is responsible for the deformed material movement, which in turn causes intermixing between the base materials [41]. FSW is capable to operate

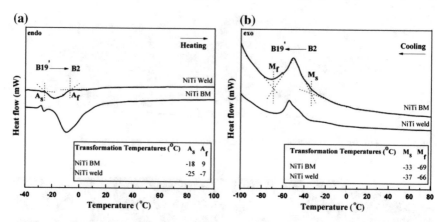

Fig. 3.10 Phase transformation behavior of FSW welded NiTi: **a** heating cycle and **b** cooling cycle
[43], with kind permission from Elsevier

for different range of materials such as aluminum, copper, titanium, steel, plastics, composites, and dissimilar materials [40–42]. Recently, the FSW is approached for shape memory alloys of NiTi material by Prabu et al. [43]. The non-consumable FSW tool material is of Densimet tungsten-based alloy that used to obtain 1.2 mm thick NiTi material. This tool material is reported as suitable material for NiTi shape memory alloys as it is not causing any tool wear. FSW forms different microstructures such as stir zone, thermo-mechanically affected zone, heat-affected zone, and base material for the most of the investigated materials. Nevertheless, the FSW of NiTi has no distinct interface observed for these different microstructures. Dynamic recrystallization is obtained by FSW in NiTi material because of high temperature and deformation in solid state. FSW of NiTi has retained the shape memory behavior relative to base material without any substantial change in transformation temperatures. However, there is minor change in transformation temperature reported as shown in Fig. 3.10, which is due to the presence of precipitates, grain sizes, dislocations, thermal stress, and detwinning. In case of FSW, detwinning and grain refinement are strong reasons proposed for the change in transformation temperatures. The weld zone obtained by FSW has retained austenitic and martensitic phases without any change relative to the NiTi base material. As mentioned above, formation of IMCs phases such as NiTi2 and TiNi3 in the weld zone is ordinary with conventional welding processes. Besides, the formation of these phases is absent in the weld obtained through FSW process. FSW has enhanced the yield strength of the weld relative to the base material due to grain refinement. The microhardness is slightly reduced at the weld zone of NiTi friction stir welds. It can be stated that FSW has high potential to opt for shape memory alloys due to its aforementioned process capabilities.

Oliveira et al. [44] performed FSW assisted by electric current on Al/NiTi composite material. Different transformation temperature of NiTi is exhibited at the weld area. At room temperature, martensite and austenite both of the phases are presented after welding whereas fully austenitic NiTi is reported before welding. This article

has proved ability of FSW to form Al/NiTi composites which will increase number of applications of this dissimilar combination.

3.4.3 Ultrasonic Welding

Ultrasonic welding is a type of solid state welding that uses high-frequency mechanical vibrations and static compressive load to generate relative motion between two faying surfaces. This friction causes heat and subsequently leads to the plastic deformation that ultimately forms joints [8]. In the case of ultrasonic welding, workpiece material hardness, and thickness are important process parameters that decides energy required to produce welds [1, 8]. Process parameters such as contact force, amplitude of ultrasonic wave, and time required to obtain weld are additional that affects formation of welds and weld properties [8]. Ultrasonic welding is most suitable for the thin parts and small size job. Ultrasonic welding of shape memory alloys is not so popular as laser welding. So far, a welding by ultrasonic welding process is not discovered for the shape memory alloys [2]. However, application of ultrasonic waves for different applications of the shape memory alloys is studied. Kong et al. [45] have performed ultrasonic technique to embedded fibers of shape memory alloys within the aluminum matrix. They found that embedding NiTi shape memory alloys in Al matrix via ultrasonic welding have improved bonding without affecting the functional properties of the alloys. It is recommended that low pressure and low amplitude of oscillation is required, where the material degradation is not observed. Additionally, the bonding is reported as stronger if the temperature is increased to about 300 °C, which is 25% of the melting temperature of the NiTi shape memory alloy.

3.4.4 Explosive Welding

Explosive welding is also a type of solid-state welding technique that requires detonation with the help of suitable chemical explosives in order to obtain extremely high velocity of workpiece. This leads to the deformation of workpiece material because of high impact energy and subsequently accomplishes the weld. In case of explosive welding, the explosive detonator is applied on the surface of one of the workpieces. The blast through detonator generates a very high velocity that in turn causes collision with another workpiece. This phenomenon leads to the plastic deformation and subsequently forms joint. Explosive welding is reported as the best technique to obtain dissimilar joints [8]. Among all solid-state welding processes, the explosive welding is most utilized process for shape memory alloys [2]. Initially, the explosive welding is explored for dissimilar welding of shape memory alloys to another material. Most commonly used shape memory alloy of NiTi is welded with steel using explosive welding, in order to take advantages such as high resistance of NiTi and cavitation erosion and structural strength of steel [2]. Explosive welding of austenitic NiTi to low carbon steel is performed by Zimmerly et al. [46]. The NiTi–carbon steel weld interface is analysed by X-ray diffraction method and identified that low-intensity martensitic peaks are reported due to the formation of shock-induced martensite. The cavitation erosion property of NiTi weld interface is slightly decreased relative to the base material, which is there because of martensitic phase presence. Loss of this erosion resistance to cavitation during welding is biggest challenge considered. Nevertheless, it is suggested that, the austenitic interface is required to have better resistance to the cavitation erosion. Hence, the post weld heat treatment is recommended to retain resistance to cavitation erosion. The failure during the tensile lap shear strength test is reported either from NiTi base material or low carbon steel while never from the interface between NiTi and low carbon steel. Richman et al. [47] conducted a similar type of study with martensitic NiTi to low carbon steel and added some new results. They also mentioned that austenitic NiTi is required to subject for post weld heat treatment in order to have erosion resistance that is lost during explosive welding. Explosive welding is also performed to obtain dissimilar NiTi to stainless steel composites. The martensitic transformation is depressed due to plastic deformation induced during explosive welding. Although, application of post weld heat treatment can recover this transformation. It is reported that no formation of brittle IMCs are observed with explosive welding of NiTi to stainless steel. In case of two-way shape memory effect of explosive welded dissimilar NiTi to stainless steel welds, the recoverable strain is exceeded because of the action of steel elastic layer. This dissimilar combination is found as perfectly suitable for actuator due to its functional properties. Attempts of NiTi to NiTi joints are also conducted with explosive welding [48–52]. Explosive welding has proved its ability to form NiTi–NiTi joints have different transformation temperatures and proportions of chemical compositions. There is a change in transformation temperature observed after the welding. Post weld heat treatment is required to obtain nearly same transformation temperature of the weld zone. However, proper selection of heat treatment temperature is critical.

3.5   Adhesive Bonding

Adhesive bonding is a type of joining technique, which uses an intermediate layer of adhesives to obtain a bond. Adhesive bonding is better than the mechanical joints and weaker than the welded joints [3]. Process parameters such as adhesive material, coating thickness of adhesive, temperature of bonding, processing time, chamber pressure, and tool pressure are important. There are some articles available, which talk about adhesive bonding of shape memory alloys. Different adhesive materials such as cyanoacrylates, epoxies, etc., can be applied to the shape memory alloys depending on different service environment and degradation susceptibility. It is reported that adhesive bonding requires surface pretreatment procedure in order to have better bonding between adhesives and shape memory alloys. Different surface treatments

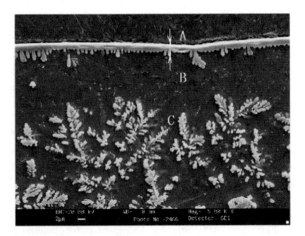

Fig. 3.11 Microstructure of laser gas nitride NiTi material (A-B: TiN, C: TiN dendrites + NiTi matrix) [53], with kind permission from Elsevier

such as acid etching, polymer coating, and sandblasting techniques are applied to enhance the adhesion between NiTi wires and epoxy matrix. Amon all these surface treatments, sandblasting is reported as most efficient method. The chemical etching is disregarded here because of significant material loss. Besides, Man and Zhao [53] studied superficial texture of NiTi to improve adhesion properties of the surfaces. The irradiation by laser is applied on NiTi plates to obtain a joint and highly pure argon is discharged at the same time on the molten pool. The chemical etching is performed to see a three-dimensional structure of TiN dendrites in a NiTi matrix structure as can be observed in Fig. 3.11. It is proposed that this type of structure is generated due to different chemical etching rate. TiN phase is transferred at the bottom of the weld due to Maragoni effect that in turn causes formation of dendritic structure. It is reported that the adhesive strength can be enhanced up to 150% than the sandblast etched samples, by providing appropriate etching time. This is due to more protruding dendrites obtained after long etching [54].

3.6 Brazing and Soldering

Brazing and soldering are types of joining processes that operate with low melting temperature filler wire than the base material. In case of brazing, generally 450 °C or lower than that is used while more than 450 °C is used in soldering [2, 8]. In these processes, the bonding is obtained at a short distance by creating diffusion as the base material is wetted by melted filler wire. Various factors such as selection of filler material, mechanical properties, wettability, and formation of IMCs must be considered for brazing and soldering.

Brazing and soldering processes are successfully applied to shape memory alloys and dissimilar materials. Reports on soldering of shape memory alloys are very limited. Dissimilar systems of NiTi-stainless steel, NiTI-titanium, NiTi-NiAl, etc., are investigated for brazing [2]. Different sources of energy are utilized to melt the filler material such as laser, electric furnace, microwave, and oxyacetylene flame. Different filler wires such as copper, titanium (Ti), niobium (Nb), silver, Ag–Cu–Ti, Ag–Ti, and other Ag-based filler wires are used to obtain brazing of NiTi base material [2]. Yang et al. [55] reported better functional properties of brazed NiTi joints using Cu as filler material for infrared laser brazing. It is reported that full recovery of shape memory effects observed with Cu filler material whereas Ti-based Ti–15Cu–15Ni filler material joints are failed in bend test due to higher brittle joints. The authors of [56–59] performed vacuum brazing on NiTi base material using Nb as filler material. They reported that brazing time significantly influences consumption of Nb and pro-eutectic of NiTi. They also mentioned enhancement in mechanical properties such as relative density of 5%, super elasticity of joints and recovery of 50% compressive strains upon unloading. Nb filler material is also mentioned by Wang et al. [60] for their work of furnace brazing of NiTi base material. They reported the eutectic reaction is generated between NiTi and Nb to braze NiTi wires with Nb powders deposited on the contact areas of the wires. Shiue and Wu [61] investigated infrared laser brazing to join equiatomic NiTi base material using Ag–Cu–Ti filler material. The shape memory recovery is influenced by Ag-rich, Cu-rich, and Ti-rich phases. It is mentioned that presence of higher Ti-rich phase is required to have superior shape memory recovery. However, minor difference in the shape memory effect is reported due to variations in microstructures. van der Eijk et al. [62] carried out brazing of NiTi with filler wires of Ag–Ti and Ag–Cu–Ti in a microwave. They reported a significant change in phases.

Dissimilar materials joining of shape memory alloys with other materials are studied for brazing and soldering. Gale and Guan [63] conducted studies on NiTi to NiAl joining by furnace brazing with the help of Cu filler material. After brazing, no significant changes of microstructure reported. However, minor change in transformation temperature is observed. Different authors have investigated NiTi to stainless steel joining system with brazing and soldering [2]. It is summarized that Ag-based filler wire is best suited for this combination. Various phases such as Ag, AgZn, Ag_3Sn, and Cu_5Zn_8 are reported inside the brazing of NiTi–stainless steel interface [2, 3]. Different elements of Ag, Cu, Sn, and Zn are diffused into these two base materials which is because of the formation of a diffusion reaction layer [2]. At the same time, elements such as Ni and Ti of NiTi base material, and Fe, Cr, and Ni of stainless steel are diffused into the brazed alloy. Besides, the shape memory effect is deteriorated at the thermally affected NiTi region [2, 3]. In case of laser autogenous brazing of dissimilar NiTi-stainless steel system, the laser is positioned either on stainless steel side or on NiTi side, and then moved toward the interface side [64, 65]. There is no filler wire applied for this type of system. By the application of this technique, the tensile strength is enhanced up to 500 MPa. However, superelastic behavior is not reported during tensile strength. Brazing of NiTi to Ni-based superalloy is attempted in microwave using Ag–Cu filler wire [62]. Complex microstructure is reported due to multi-phase diffusion. Negative results with defect such as porosity are reported and suggested that Ag–Cu filler wire is not recommended for dissimilar NiTi to Hastalloy system [66]. Ag–Cu filler wire is adopted for another dissimilar system of NiTi to NiTiNb materials and observed wide recovery of temperature such as −60 to +45°C.

3.7 Summary

Different welding processes such as tungsten inert gas welding, plasma welding, laser beam welding, electron beam welding, resistance welding, friction stir welding, friction welding, explosive welding, ultrasonic welding, diffusion bonding, adhesive bonding, brazing and soldering are discussed for process capabilities and challenges to obtain shape memory welds. Phase transformation temperature, shape memory effect of welds, microstructure of weld, formation of IMCs and weld properties are summarized with respect to these welding processes. It is reported that laser welding technique is the most investigated process among available welding processes due to its advantages of better process control, low heat input, high density, ability to reproduce, low HAZ, and monochromatic nature. Welding of shape memory alloy with another material is also discussed for the explored welding processes. Laser welding and solid-state welding processes are mostly applied to obtain dissimilar joint of shape memory alloy with another material. Solid state welding processes can be further developed and studied considering its ability to retain shape memory effect and no formation of detrimental IMCs in the weld zone. In some cases, post weld treatments are required in order to retain the shape memory effect of the weld. Considering the existing and potential applications of shape memory alloys, there is a need to develop welding and joining processes for different shape memory alloys. Majority of the studies are focused on NiTi-type shape memory alloy, which can be extended for the other shape memory alloys.

References

1. K. Mehta, Advanced joining and welding techniques: an overview, in *Advanced Manufacturing Technologies* (Springer International Publishing, 2017), pp. 101–136. https://link.springer.com/chapter/10.1007/978-3-319-56099-1_5
2. J.P. Oliveira, R.M. Miranda, F.B. Fernandes, Welding and joining of NiTi shape memory alloys: a review. Prog. Mater Sci. **88**, 412–466 (2017)
3. O. Akselsen, Joining of shape memory alloys, in *Shape Memory Alloys* (InTech, 2010), pp. 183–210
4. A. Ikai, K. Kimura, H. Tobushi, TIG welding and shape memory effect of TiNi shape memory alloy. J. Intell. Mater. Syst. Struct. **7**(6), 646–655 (1996)
5. J.P. Oliveira, D. Barbosa, F.B. Fernandes, R.M. Miranda, Tungsten inert gas (TIG) welding of Ni-rich NiTi plates: functional behavior. Smart Mater. Struct. **25**(3), 03LT01 (2016)

6. G. Fox, R. Hahnlen, M.J. Dapino, Fusion welding of nickel–titanium and 304 stainless steel tubes: Part II: tungsten inert gas welding. J. Intell. Mater. Syst. Struct. **24**(8), 962–972 (2013)
7. S.X. Lue, Z.L. Yang, H.G. Dong, Welding of shape memory alloy to stainless steel for medical occluder. Trans. Nonferrous Met. Soc. China **23**(1), 156–160 (2013)
8. J. Norrish, *Advanced Welding Processes* (Institute of Physics, 1992)
9. C. van der Eijk, H. Fostervoll, Z.K. Sallom, O.M. Akselsen, Plasma welding of NiTi to NiTi, stainless steel and hastelloy C276, in *Proceedings of the ASM Materials Solutions Conference* (2003), pp. 125–129
10. G.J. Julien, A. Sickinger, G.A. Hislop, U.S. Patent No. 6,043,451. (U.S. Patent and Trademark Office, Washington, DC, 2000)
11. S. Ozel, B. Kurt, I. Somunkiran, N. Orhan, Microstructural characteristic of NiTi coating on stainless steel by plasma transferred arc process. Surf. Coat. Technol. **202**(15), 3633–3637 (2008)
12. J.P.D.S. Oliveira, Laser welding of shape memory alloys, Doctoral thesis (2016), pp. 1–187
13. J.P. Oliveira, B. Panton, Z. Zeng, T. Omori, Y. Zhou, R.M. Miranda, F.B. Fernandes, Laser welded superelastic Cu–Al–Mn shape memory alloy wires. Mater. Des. **90**, 122–128 (2016)
14. A. Falvo, F.M. Furgiuele, C. Maletta, Laser welding of a NiTi alloy: mechanical and shape memory behaviour. Mater. Sci. Eng., A **412**(1), 235–240 (2005)
15. C.W. Chan, H.C. Man, Laser welding of thin foil nickel–titanium shape memory alloy. Opt. Lasers Eng. **49**(1), 121–126 (2011)
16. C.W. Chan, H.C. Man, T.M. Yue, Effect of postweld heat treatment on the microstructure and cyclic deformation behavior of laser-welded NiTi-shape memory wires. Metall. Mater. Trans. A **43**(6), 1956–1965 (2012)
17. C.W. Chan, H.C. Man, F.T. Cheng, Fatigue behavior of laser-welded NiTi wires in small-strain cyclic bending. Mater. Sci. Eng., A **559**, 407–415 (2013)
18. J.P. Oliveira, F.B. Fernandes, N. Schell, R.M. Miranda, Shape memory effect of laser welded NiTi plates. Funct. Mater. Lett. **8**(06), 1550069 (2015)
19. G.R. Mirshekari, A. Kermanpur, A. Saatchi, S.K. Sadrnezhaad, A.P. Soleymani, Microstructure, cyclic deformation and corrosion behavior of laser welded NiTi shape memory wires. J. Mater. Eng. Perform. **24**(9), 3356–3364 (2015)
20. C.W. Chan, H.C. Man, T.M. Yue, Parameter optimization for laser welding of NiTi wires by the taguchi method. Lasers Eng. (Old City Publishing) **30** (2015)
21. A. Bahador, S.N. Saud, E. Hamzah, T. Abubakar, F. Yusof, M.K. Ibrahim, Nd: YAG laser welding of Ti-27 at.% Nb shape memory alloys. Weld. World **60**(6), 1133–1139 (2016)
22. P. Sathiya, T. Ramesh, Experimental investigation and characterization of laser welded NiTinol shape memory alloys. J. Manuf. Processes **25**, 253–261 (2017)
23. C.W. Chan, H.C. Man, T.M. Yue, Susceptibility to environmentally induced cracking of laser-welded NiTi wires in Hanks' solution at open-circuit potential. Mater. Sci. Eng., A **544**, 38–47 (2012)
24. M. Mehrpouya, A. Gisario, M. Elahinia, Laser welding of NiTi shape memory alloy: a review. J. Manuf. Processes **31**, 162–186 (2018)
25. Z. Zeng, J.P. Oliveira, M. Yang, D. Song, B. Peng, Functional fatigue behavior of NiTi-Cu dissimilar laser welds. Mater. Des. **114**, 282–287 (2017)
26. J.P. Oliveira, Z. Zeng, C. Andrei, F.B. Fernandes, R.M. Miranda, A.J. Ramirez, T. Omori, N. Zhou, Dissimilar laser welding of superelastic NiTi and CuAlMn shape memory alloys. Mater. Des. **128**, 166–175 (2017)
27. P. Schlossmacher, T. Haas, A. Schüssler, Laser-welding of a Ni-rich TiNi shape memory alloy: mechanical behavior. J. Phys. IV **7**(C5), C5–251 (1997)
28. Y.G. Song, W.S. Li, L. Li, Y.F. Zheng, The influence of laser welding parameters on the microstructure and mechanical property of the as-jointed NiTi alloy wires. Mater. Lett. **62**(15), 2325–2328 (2008)
29. A. Falvo, F.M. Furgiuele, C. Maletta, Functional behaviour of a NiTi-welded joint: two-way shape memory effect. Mater. Sci. Eng., A **481**, 647–650 (2008)

30. L.A. Vieira, F.B. Fernandes, R.M. Miranda, R.J.C. Silva, L. Quintino, A. Cuesta, J.L. Ocaña, Mechanical behaviour of Nd: YAG laser welded superelastic NiTi. Mater. Sci. Eng., A **528**(16), 5560–5565 (2011)
31. B. Panton, A. Pequegnat, Y.N. Zhou, Dissimilar laser joining of NiTi SMA and MP35N wires. Metall. Mater. Trans. A **45**(8), 3533–3544 (2014)
32. D. Yang, H.C. Jiang, M.J. Zhao, L.J. Rong, Microstructure and mechanical behaviors of electron beam welded NiTi shape memory alloys. Mater. Des. **57**, 21–25 (2014)
33. D. Yang, H.C. Jiang, M.J. Zhao, L.J. Rong, Effect of post-weld annealing on microstructure and properties of NiTi welding joints. Mater. Res. Innovations **18**(sup4), S4–588 (2014)
34. B. Tam, A. Pequegnat, M.I. Khan, Y. Zhou, Resistance microwelding of Ti-55.8 wt pct Ni nitinol wires and the effects of pseudoelasticity. Metall. Mater. Trans. A **43**(8), 2969–2978 (2012)
35. K. Mehta, M. Gupta, P. Sharma, Nano-machining, nano-joining, and nano-welding, in *Micro and Precision Manufacturing* (Springer, Cham, 2018), pp. 71–86. https://link.springer.com/chapter/10.1007/978-3-319-68801-5_4
36. V. Delobelle, P. Delobelle, Y. Liu, D. Favier, H. Louche, Resistance welding of NiTi shape memory alloy tubes. J. Mater. Process. Technol. **213**(7), 1139–1145 (2013)
37. T. Shinoda, T. Tsuchiya, H. Takahashi, Friction welding of shape memory alloy. Weld. Int. **6**(1), 20–25 (1992)
38. T. Shinoda, T. Owa, V. Magula, Microstructural analysis of friction welded joints in TiNi alloy. Weld. Int. **13**(3), 180–185 (1999)
39. S. Fukumoto, T. Inoue, S. Mizuno, K. Okita, T. Tomita, A. Yamamoto, Friction welding of TiNi alloy to stainless steel using Ni interlayer. Sci. Technol. Weld. Joining **15**(2), 124–130 (2010)
40. K.P. Mehta, V.J. Badheka, A review on dissimilar friction stir welding of copper to aluminum: process, properties, and variants. Mater. Manuf. Processes **31**(3), 233–254 (2016)
41. K.P. Mehta, V.J. Badheka, Influence of tool design and process parameters on dissimilar friction stir welding of copper to AA6061-T651 joints. Int. J. Adv. Manuf. Technol. **80**(9–12), 2073–2082 (2015)
42. K.P. Mehta, V.J. Badheka, Hybrid approaches of assisted heating and cooling for friction stir welding of copper to aluminum joints. J. Mater. Process. Technol. **239**, 336–345 (2017)
43. S.M. Prabu, H.C. Madhu, C.S. Perugu, K. Akash, P.A. Kumar, S.V. Kailas, M. Anbarasu, I.A. Palani, Microstructure, mechanical properties and shape memory behaviour of friction stir welded nitinol. Mater. Sci. Eng., A **693**, 233–236 (2017)
44. J.P. Oliveira, J.F. Duarte, P. Inácio, N. Schell, R.M. Miranda, T.G. Santos, Production of Al/NiTi composites by friction stir welding assisted by electrical current. Mater. Des. **113**, 311–318 (2017)
45. C.Y. Kong, R.C. Soar, P.M. Dickens, Ultrasonic consolidation for embedding SMA fibres within aluminium matrices. Compos. Struct. **66**(1), 421–427 (2004)
46. C.A. Zimmerly, O.T. Inal, R.H. Richman, Explosive welding of a near-equiatomic nickel-titanium alloy to low-carbon steel. Mater. Sci. Eng., A **188**(1–2), 251–254 (1994)
47. R.H. Richman, A.S. Rao, D. Kung, Cavitation erosion of NiTi explosively welded to steel. Wear **181**, 80–85 (1995)
48. T. Xing, Y. Zheng, L. Cui, Transformation and damping characteristics of NiTi/NiTi alloys synthesized by explosive welding. Mater. Trans. **47**(3), 658–660 (2006)
49. J. Li, Y. Zheng, L. Cui, Transformation characteristics of TiNi/TiNi alloys synthesized by explosive welding. Front. Mater. Sci. Chin. **1**(4), 351–355 (2007)
50. L. Juntao, Z. Yanjun, C. Lishan, Effects of severe plastic deformation and heat treatment on transformation behavior of explosively welded duplex TiNi–TiNi. Pet. Sci. **4**(4), 107–112 (2007)
51. Z. Yan, L.S. Cui, Y.J. Zheng, Microstructure and martensitic transformation behaviors of explosively welded NiTi/NiTi laminates. Chin. J. Aeronaut. **20**(2), 168–171 (2007)
52. T.Y. Xing, Y.J. Zheng, L.S. Cui, X.J. Mi, Influence of aging on damping behavior of TiNi/TiNi alloys synthesized by explosive welding. Trans. Nonferrous Met. Soc. China **19**(6), 1470–1473 (2009)

53. H.C. Man, N.Q. Zhao, Enhancing the adhesive bonding strength of NiTi shape memory alloys by laser gas nitriding and selective etching. Appl. Surf. Sci. **253**(3), 1595–1600 (2006)
54. F. Niccoli, M. Alfano, L. Bruno, F. Furgiuele, C. Maletta, Mechanical and functional properties of nickel titanium adhesively bonded joints. J. Mater. Eng. Perform. **23**(7), 2385–2390 (2014)
55. T.Y. Yang, R.K. Shiue, S.K. Wu, Infrared brazing of Ti 50 Ni 50 shape memory alloy using pure Cu and Ti–15Cu–15Ni foils. Intermetallics **12**(12), 1285–1292 (2004)
56. D.S. Grummon, J.A. Shaw, J. Foltz, Fabrication of cellular shape memory alloy materials by reactive eutectic brazing using niobium. Mater. Sci. Eng., A **438**, 1113–1118 (2006)
57. D. Grummon, K.B. Low, J. Foltz, J. Shaw, A new method for brazing nitinol based on the quasi-binary TiNi-Nb system, in *48th AIAA/ASME/ASCE/AHS/ASC Structures, Structural Dynamics, and Materials Conference* (American Institute of Aeronautics and Astonautics, 2007), p. 1741
58. J.A. Shaw, D.S. Grummon, J. Foltz, Superelastic NiTi honeycombs: fabrication and experiments. Smart Mater. Struct. **16**(1), S170 (2007)
59. J.A. Shaw, C. Churchill, N. Triantafyllidis, P. Michailidis, D. Grummon, J. Foltz, Shape memory alloy honeycombs: experiments and simulation, in *Proceedings of the AIAA/ASME/ASCE/AHS/ASC Structures, Structural Dynamics and Materials Conference*, vol. 1 (2007), pp. 428–436
60. L. Wang, C. Wang, D.C. Dunand, Microstructure and strength of NiTi-Nb eutectic braze joining NiTi wires. Metall. Mater. Trans. A **46**(4), 1433–1436 (2015)
61. R.H. Shiue, S.K. Wu, Infrared brazing of Ti 50 Ni 50 shape memory alloy using two Ag–Cu–Ti active braze alloys. Intermetallics **14**(6), 630–638 (2006)
62. C. van der Eijk, Z.K. Sallom, O.M. Akselsen, Microwave brazing of NiTi shape memory alloy with Ag–Ti and Ag–Cu–Ti alloys. Scripta Mater. **58**(9), 779–781 (2008)
63. W.F. Gale, Y. Guan, Microstructural development in copper-interlayer transient liquid phase bonds between martensitic NiAl and NiTi. J. Mater. Sci. **32**(2), 357–364 (1997)
64. M.G. Li, D.Q. Sun, X.M. Qiu, D.X. Sun, S.Q. Yin, Effects of laser brazing parameters on microstructure and properties of TiNi shape memory alloy and stainless steel joint. Mater. Sci. Eng., A **424**(1), 17–22 (2006)
65. M.G. Li, D.Q. Sun, X.M. Qiu, S.Q. Yin, Corrosion behavior of the laser-brazed joint of TiNi shape memory alloy and stainless steel in artificial saliva. Mater. Sci. Eng., A **441**(1), 271–277 (2006)
66. X.K. Zhao, L. Lan, H.B. Sun, J.H. Huang, H. Zhang, Preparation of NiTi/NiTiNb laminated alloys by vacuum brazing, in *Advanced Materials Research*, vol. 97 (Trans Tech Publications, 2010), pp. 1653–1656

Chapter 4
Processing of Shape Memory Alloys

4.1 Introduction

Processing of material is defined as a series of operations performed to fabricate the raw material into nearly finished product. Different manufacturing processes are followed to obtain finished product. There are numbers of processing techniques performed for the shape memory alloys. Retention of shape memory effect, mechanical properties variations, microstructural changes, and formation of intermetallic compounds (IMCs) and precipitates are issues associated with processing of shape memory alloys as thermal and or mechanical treatments affect chemical reactions and phase transformation of shape memory alloys [1–3]. Process parameters affecting these issues are discussed in this chapter under a different classification of processing techniques such as powder metallurgy processing, additive processing, thermomechanical processing, and mechanical processing [1–3]. Further classifications of these topics are presented in Fig. 4.1 that are discussed individually in subsequent sections. Conventional and advanced processing techniques have been attempted to shape memory alloys are considered in this chapter. Shape memory alloys such as NiTi, FeMnSi, Fe–Mn–Si–Cr–Ni Cu–Zn–Al, Cu–Al–Ni, and Ni–Fe–Ga are reported with different processing techniques [1–7].

4.2 Powder Metallurgy Processing

Powder metallurgy processing is a conventional method of material processing in which the metal powder is processed to obtain final product without any material removal. Various powder metallurgy techniques such as sintering, hot isostatic pressing, metal injection molding, self-propagating high-temperature synthesis, and spark plasma sintering are reported for different shape memory alloys in the large number of literatures. Shape memory alloys of NiTi, Cu–Al–Ni–Mn, Cu–Al–Ni, and

© The Author(s), under exclusive licence to Springer Nature Switzerland AG 2019
K. Mehta and K. Gupta, *Fabrication and Processing of Shape Memory Alloys*,
Manufacturing and Surface Engineering, https://doi.org/10.1007/978-3-319-99307-2_4

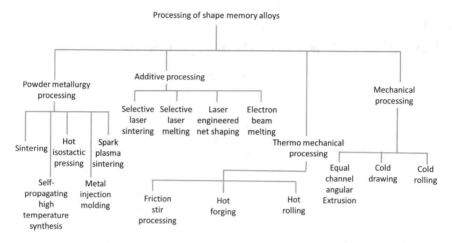

Fig. 4.1 Classification of processing of shape memory alloys

Fe–Mn–Si–Cr–Ni are processed and investigated by different aforementioned powder metallurgy techniques [1–12]. However, the majority of reports are on NiTi shape memory alloy considering materials acceptability for various useful applications. Each of the abovementioned processes are discussed one by one as under.

4.2.1 Conventional Sintering

Sintering is the process of fabricating solid mass of material by compacting and forming with the application of heat or pressure without melting it to the point of liquification. Conventional sintering is majorly applied to NiTi shape memory alloys [12, 13]. Elemental powder of Ni and Ti is blended and processed through conventional sintering in order to make shape memory alloy. The problems such as homogenization and obtaining high density are associated with conventional sintering of Ni and Ti elemental powders with equiatomic blends. One of the major problems in fabrication of NiTi material by conventional sintering is formation of porosity. There are few factors such as applied pressure, unbalance in diffusion of Ni into Ti and Ti into Ni, shrink during sintering, and capillary forces are identified that governs formation of porosity in conventional sintering of NiTi shape memory alloy [12, 13]. Homogeneous NiTi intermetallic is required in order to attain the maximum shape memory transformation enthalpy in case of conventional sintering processed by Ni and Ti elemental powders [13]. NiTi processed by conventional sintering is compatible with commercially produced wrought NiTi material in terms of shape memory effect [13]. Density of NiTi processed by conventional sintering is important in order to exhibit the shape memory effect up to a meaningful extent [12]. However, they concluded that porous $Ti_{49}Ni_{51}$ is obtained through normal sintering that can

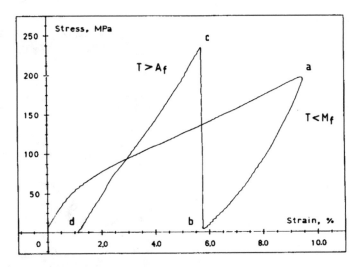

Fig. 4.2 Shape memory effect of porous $Ti_{49}Ni_{51}$ is obtained through normal sintering [12], with kind permission from Elsevier

exhibit shape memory effect up to the meaningful level as can be seen from Fig. 4.2 and they claimed that it is comparable with same cast shape memory alloy.

Apart from NiTi shape memory alloy, other materials such as Fe–Mn–Si–Cr–Ni and Cu–Al–Ni are attempted under the process of conventional sintering [6, 7]. Conventional sintering of Cu–Al–Ni is investigated very first and reported successful martensitic transformation. It is reported that superior mechanical properties can be obtained using powder metallurgy compare to the conventional casting process. However, no major changes in the shape memory properties are mentioned [6]. Processing of Fe-based shape memory alloy by conventional sintering affects pseudoelasticity, ductility, and microstructures significantly with the application of different strain values [7]. The thermally induced reversion of stress-induced martensitic structure is reported in case of Fe–Mn–Si–Cr–Ni shape memory alloy processed by conventional sintering process [7].

4.2.2   Hot Isostatic Pressing

Hot isostatic pressing (HIP) is also known as pressure enhanced sintering that increases density and reduces porosity of the material, which is the major drawback of the conventional sintering process in case of shape memory alloys [1]. Metallic powder is subjected to elevated temperature and isostatic gas pressure in a high-pressure vessel as shown in Fig. 4.3. The elemental powder mixture is encapsulated in a vacuumed canister that is subjected to simultaneous isostatic pressure at elevated temperature [1]. This leads to the shape memory alloys made with 0% porosity. Apart

Fig. 4.3 Hot isostatic
pressing

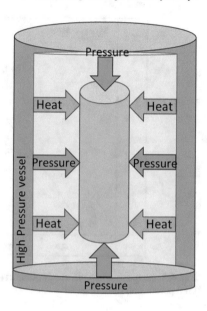

from using vacuumed canister, the argon gas can be used as the surrounding environment instead of airtight chamber. HIP technique is reported for two most widely used shape memory alloys such as NiTi and Cu–Al–Ni–Mn [14–17]. It is reported that the stress-induced martensitic transformation is observed in the entire medium at same stress level. Different regions of martensitic and austenitic structures of phase transformation of the stress–strain curve can be easily identified due to the dense structure of HIP fabricated shape memory alloys [1]. Presence of precipitates and formation of IMCs cannot be avoided in case of HIP, for example, formation of $NiTi_2$ and Ni_3Ti precipitates in case of NiTi shape memory alloys [1, 11]. The shape memory effect can be obtained with shape memory alloys processed with HIP with a control on super elasticity and porosity [1, 17]. The shape memory recovery is reported 100% under 100 times cycling for Cu–Al–Ni–Mn shape memory alloy processed by vacuum hot pressing [5].

A variant of HIP investigated for NiTi shape memory alloys is called capsule-free HIP, which produces homogeneous porous material with near spherical pores [14, 18]. This type of structure leads to acceptable pseudoelasticity as stress concentration is not occurred in near spherical pores. This has also proved that controlling the porosity of the porous shape memory alloy greatly influences the mechanical properties of the alloy [14, 18]. An example of capsule-free HIP processed NiTi having different porosity for shape memory effect is shown in Fig. 4.4.

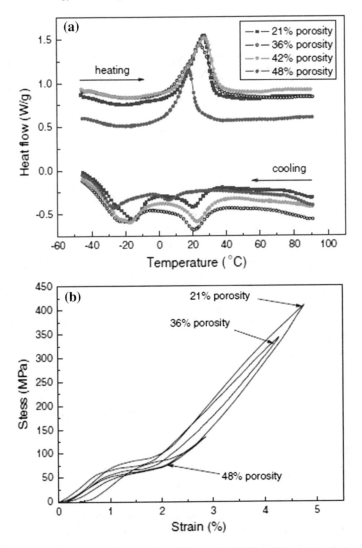

Fig. 4.4 Shape memory effect of capsule-free HIP processed NiTi **a** differential calorimetry curve and **b** compressive stress–strain curve [14], with kind permission from Elsevier

4.2.3 Metal Injection Molding

Metal injection molding is a powder metallurgy technique that works on the concept of plastic injection molding. Metal injection molding leads advantages such as geometrical precision of the parts, high production value, and low cost [1]. Metal injection molding of NiTi shape memory alloys is reported in [10, 19, 20]. The process operates with four steps of (I) feedstock fabrication, (II) injection molding, (III)

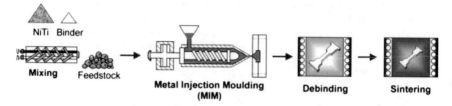

NiTi Binder

Mixing   Feedstock

Metal Injection Moulding
(MIM)

Debinding

Sintering

Fig. 4.5 Metal injection molding process [20], with kind permission from John Wiley and Sons

debinding, and (IV) sintering as shown in Fig. 4.5. The elemental powder is mixed
with a binder in the first step of feedstock fabrication. In the next step, it is injected
into the mold with specific suitable temperature and pressure. Next to it, the debinding is performed that is performed in a chemical bath with an elevated temperature
under a vacuum. In the end, the sintering is conducted at high temperature in order to
meet the density requirement of the material [1]. Metal injection molding is adopted
for NiTi material fabrication specifically for biomedical applications [10, 19, 20].
One way effect of pseudoelasticity is reported for NiTi shape memory alloy. It is
also reported that no loss of structural integrity is observed after 1.2×10^6 of load-
ing/unloading cycles in air or saline solution for NiTi fabricated by metal injection
molding. Different cyclic behavior of stress–strain curve is reported with different
temperatures [19, 20].

4.2.4 Self-propagating High-Temperature Synthesis

Self-propagating high-temperature synthesis (SHS) is a type of powder metallurgy
process in which the specimen is subjected to thermal explosion from one of its ends
[1]. This thermal explosion propagates in a specimen in a self-sustaining manner as
shown in Fig. 4.6.

NiTi shape memory alloy is fabricated by SHS process [1, 21, 22]. Exothermic
reaction between Ni and Ti is caused by thermal explosion and that consequently
leads to the fabrication of NiTi alloys. It is noted that the formation of IMCs is difficult
to control in case of shape memory alloys processed by SHS [21, 22]. Formation
of porosity is also a problem faced by researchers in case of NiTi processing by
SHS. However, synthesis process parameters of the change in the molar volume, the
combustion front thermal gradients, and the gas evolution controls the formation of
porosity. It is also mentioned that manipulation with reaction temperature along with
the addition of diluents can control the pore size in SHS process of shape memory
alloys [1, 21, 22]. In case of NiTi processing by SHS leads to the formation of
different phases such as Ti_2Ni, Ni_3Ti, and Ni_4Ti_3 that are usually reported in the
product matrix, which consequently causes the corresponding embrittlement effects
[1, 21, 22]. An example of porous NiTi processed by SHS and identified IMCs in
the microstructural examinations are shown in Fig. 4.7 from the article of [22].

Fig. 4.6 Self-propagating
high-temperature synthesis
process [1], with kind
permission from Elsevier

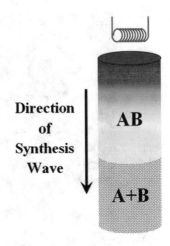

4.2.5 Spark Plasma Sintering

Spark plasma sintering (SPS) is a powder metallurgy process also known as pulsed
electric current sintering (PECS) that use to process porous shape memory alloys
[1]. In the SPS process, the prealloyed powders are subjected to press in a graphite
die. After that, the step current is supplied to the compact while the pulsed current
provides high energy to impact. This results in fabrication by joint formation between
particles at relatively low pressure and short processing time than other sintering processes. Schematic of SPS process is shown in Fig. 4.8 [1]. Shape memory alloys of
NiTi, Ni–Mn–Ga, Ni–Co–Mn–Si, Cu–Al–Ni, and Cu–Al–Mn are processed by SPS
technique [1, 23–27]. It is mentioned that the decomposition process in Cu–Al–Ni
shape memory alloy cannot be avoided in SPS processing even with the application
with short processing time. It is also reported that the martensitic transformation is
noted in most of the sintered compact during the cooling in spark plasma appara-
tus with the observed phase of Cu_9Al_4 [25]. Shape memory effect is reported for
aforementioned alloys fabricated through SPS [1, 23–27].

4.3 Additive Processing

Additive processing or additive manufacturing is an advanced manufacturing classification of processing material by adding successive layers of the material to prepare
a final part/product. Addition of metal layer is done with the help of metallic powder
that is processed with the help of well-controlled processing technique [1]. It is also
popular as rapid prototyping or rapid manufacturing. Additive processing is capable to produce intricate parts and geometries without any material removal. It is a
sustainable manufacturing approach due to having advantages of cost-effective pro-

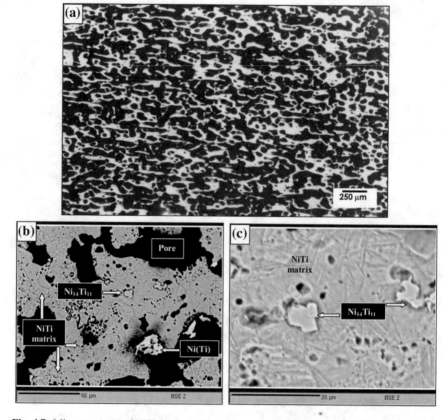

Fig. 4.7 Microstructures of NiTi fabricated by self-propagating high-temperature synthesis showing **a** porous microstructure, **b** formation of IMCs, **c** IMCs in NiTi matrix [22], with kind permission from Elsevier

cessing, materials savings, excellent energy efficiency and environmentally friendly processing [1]. Various additive processing techniques such as selective laser sintering, selective laser melting, laser engineered net shaping, and electron beam melting are reported for shape memory alloys and discussed below in subsections. These processes are having different parameters considering various categories as presented in Table. 4.1 that govern properties of shape memory alloys.

4.3.1 Selective Laser Sintering and Selective Laser Melting

Selective laser sintering (SLS) is type of additive manufacturing techniques, which uses computer-controlled, high-power laser to fuse small powdered materials that is arranged to obtain direct three-dimensional (3D) shape as shown in Fig. 4.9.

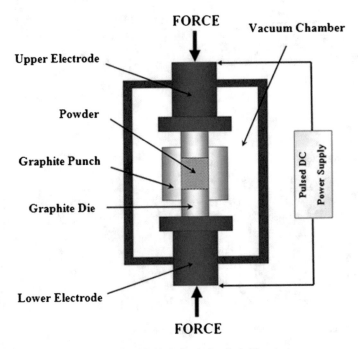

Fig. 4.8 Spark plasma sintering [1], with kind permission from Elsevier

Table 4.1 Summary of process parameters for additive processing

Sl. No.	Category	Parameters
1	Beam related parameters (Laser/Electron beam)	Beam power, Spot size, Pulse duration, Pulse frequency, Wavelength, Bandwidth
2	Scanning parameters	Scan speed, Spacing, and Scan pattern
3	Powder-related parameters	Particle shape, Particle size, Powder density, Powder distribution, Powder layer thickness
4	Temperature-related parameters	Powder bed temperature, Powder feeder temperature, Temperature uniformity

Preheating is applied to the metallic powders in most of the cases to minimize the high power requirements of laser and that also prevent warping and shrinkage. Laser scanning technique impinges on 0.1 thick powder layer that uses an enclosed chamber filled with nitrogen gas so that oxidation can be minimized. Subsequently, new layer is applied and process is repeated. Shape memory alloys such as NiTi, Cu–Al–Ni–Mn, Cu–Al–Ni–Mn–Zr, Fe–Mn–Al–Ni, and Al–Fe–V–Si are reported as fabricated by

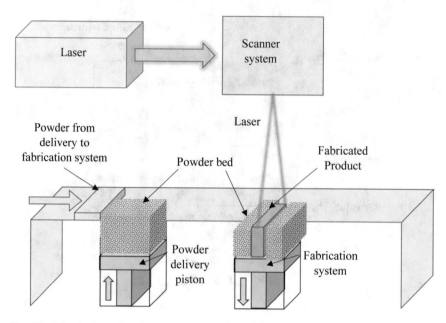

Fig. 4.9 Selective laser sintering and selective laser melting

SLM and selective laser melting (SLM) techniques in different literatures [26–36]. SLM is a successful fabrication process to obtain porous shape memory alloys with better shape memory properties, which works on similar process principle as shown in Fig. 4.9 for SLS technique. The major difference in the SLS and SLM process is the use of different type laser such as CO_2 is reported in SLS and Nd:YAG type is reported in SLM. It is observed that the application of laser with better wavelengths leads to the good absorptivity of metal powders that is reported as superior feature of SLM technique [1]. SLM with fiber laser is cheaper in terms of maintenance and setup, better energy efficiency, compactness, and beam quality. SLM has advantage of scanning pattern of f-theta lenses that minimizes distortion during scanning process [1].

Fabrication of NiTi material from SLS technique leads to the formation of phases such as Ti_3Ni, Ti_2Ni, and NiTi with compositions ranging from 60 to 80% mass wise [1]. It is also reported that SLS and SHS techniques are combined to synthesize NiTi material, which leads to the homogeneity in porosity, chemical composition, biocompatibility, and successful martensitic transformation (in the range of 50–0 °C) [1, 36]. Other than NiTi shape memory material, SLS is capable to fabricate different shape memory alloys. The formation of porosity, precipitates, IMCs, density, transformation temperature, and shape memory properties can be governed by parameters of the SLS such as powder size, preheating, laser power, beam diameter, wavelength, scan velocity, and hatch distance [1, 26–36]. It is reported that SLM-fabricated NiTi

Fig. 4.10 Laser engineered net shaping [1], with kind permission from Elsevier

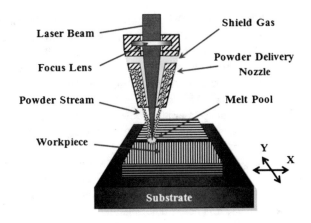

shape memory alloy has a wider temperature range for the phase transformation relative to conventionally processed NiTi [1].

A good pseudoelasticity can be produced with reversible martensitic transformation and coarse grain microstructure (having epitaxial grain growth and anisotropic microstructure) when Fe-based shape memory alloy is fabricated by SLM process [32]. Similarly, the shape memory properties can be satisfactorily obtained with Cu-based shape memory alloys fabricated by SLM process [33–35]. Process parameters of the SLM technique significantly govern porosity, phase formation, thermal stability, and mechanical properties of the shape memory alloys [35]. Formation of oxides can be avoided in case of fabrication of shape memory alloys with SLM technique [33].

Shape memory alloys such as NiTi, Cu–Al–Ni–Mn, Cu–Al–Ni–Mn–Zr, Fe–Mn–Al–Ni, and Al–Fe–V–Si are developed by SLS and SLM techniques have major applications in the area of as Micro Electro Mechanical Systems (MEMS) sensors, tissue engineering and medical and surgical equipment and implants [1, 26–36].

4.3.2 Laser Engineered Net Shaping

Laser engineered net shaping (LENS) is a type of additive processing, in which the metallic powder is injected into the molten pool as shown in Fig. 4.10 [1]. Laser head or substrate table is moved through 3D computer-aided design software that helps to deposit the material by following complex contour. Layer by layer material addition is carried out with the help of vertical head movement. The metallic powders are injected and distributed around the circumference of the head with the help of pressurized carrier gas or naturally by gravity. The shielding is usually provided with the help of an inert gas in order to prevent atmospheric contamination and reaction with oxygen [1].

LENS is successfully applied to fabricate shape memory alloys [1]. LENS can able to fabricate the products with full density and strong bond, and hence the porosity in the product of shape memory alloys is created with partial melting of metallic powder. LENS is successfully applied to obtain fully dense and chemically homogeneous NiTi equiatomic alloy while the undesirable phases are not reported [29]. It is demonstrated that formation of IMCs such as $NiTi_2$, Ni_4Ti_3, and Ni_3Ti are generally reported with fabrication of shape memory alloys. Besides, LENS process avoids formation of these IMCs [37]. It is reported that the transformation temperatures are increased in case of LENS processed NiTi shape memory material due to rapid solidification rate of LENS processing. However, it can be tailored with the help of heat treatment processes after fabrication [29]. NiTi material fabricated by LENS has many applications in implants and tissue engineering as complex parts are easy to fabricate with LENS. Nevertheless, it is noted that finishing processes are required to be performed after the fabrication of material by LENS as surface roughness is not up to the mark considering medical applications [1]. Apart from these advantages, LENS is reported limitedly in the literatures for shape memory alloys.

4.3.3 Electron Beam Melting

Electron beam melting (EBM) is a layer by layer material addition manufacturing technique using electron beam. The process works similar to SLM technique. As electron beam is involved, the process requires to operate in vacuum chamber as shown in Fig. 4.11 [1]. EBM is designed by a company called "Arcam" and developed it for orthopedic implant component manufacturing [1]. Fabrication of titanium alloys is popular with EBM technique while shape memory alloys are limitedly investigated.

It is reported that EBM technique can fabricate shape memory alloy such as NiTi with better properties relative to NiTi fabricated by vacuum induction melting (VIM). EBM produces better NiTi material in terms of homogeneous chemical compositions and reduces carbon and oxygen contaminations as compared to VIM [38, 39]. It is demonstrated that the homogeneity in chemical compositions leads to only small variation in martensitic transformation temperatures and carbon content such as 0.013 wt% as compared to the commercial products (such as 0.04–0.06 wt%) [40]. Presence of oxygen and carbon influences martensitic transformation temperatures. The presence of carbon and oxygen also affects formation of TiC and Ti_4Ni_2O, precipitates respectively in case of fabrication of NiTi by EBM technique [38]. EBM is a potential technique for the fabrication of different range of shape memory alloys.

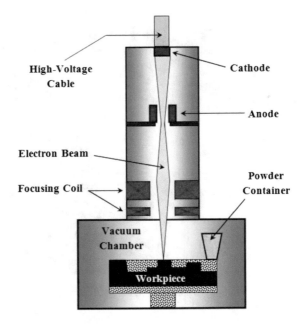

Fig. 4.11 Electron beam melting [1] with kind permission from Elsevier

4.4 Thermo-Mechanical Processing

Thermo-mechanical processing is a category of fabrication processes that undergo mechanical deformation or plastic deformation process along with thermal processing effects. Friction stir processing, hot rolling, forging and heat treatment and extrusion with heat are examples of thermo-mechanical processing.

4.4.1 Friction Stir Processing

Friction stir processing (FSP) is a type of processing techniques, which operates on the principle of solid-state processing. FSP is derived from friction stir welding that uses non-consumable rotating tool to generate frictional heat through rubbing action between workpiece and tool [41–43]. FSP process is schematically presented in Fig. 4.12. Plastic deformation is generated due to frictional heat and axial pressure. Subsequently, plastically deformed material is stirred and mixed with tool rotation and transverse movement, which leads to the microstructural and properties modifications [43].

FSP of shape memory alloys is possible with a non-consumable tool that has higher properties than the base material. FSP tool materials such as polycrystalline boron nitride, tungsten-rhenium carbide, and tool steel are reported to fabricate different range of shape memory alloys [44–47]. FSP is a technique through which the desired

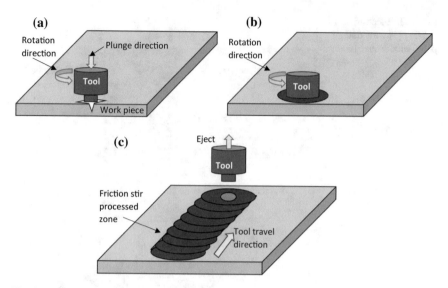

Fig. 4.12 Friction stir processing, **a** Plunge, **b** dwell and **c** processing and tool eject

processed region is produced at specific regions and depths considering FSP tool design. Mechanical properties, microstructures, and shape memory properties are tailored by FSP parameters such as rotational speed, travel speed, axial load, number of passes, tool design and geometry, and process condition [44–47]. It is observed that the shape memory capability and ductility of FSPed region of NiTi material are reduced reported by Barcellona et al. [44]. On contrary, NiTi processed by FSP leads to the increased in strength due to sever grain refinement while no decrease in ductility observed that is reported by London et al. [45]. They also reported that shape memory properties including superelasticity reported as retained.

It is well demonstrated that FSP can be utilized to fabricate metal matrix composite of shape memory alloys. Fabrication of Aluminum-NiTi composite is carried out and observed that homogeneous distribution of NiTi powder in matrix is possible with good bonding by controlling FSP process parameters. FSP fabricated composites can exhibit the similar behavior of shape memory effect as received [46, 47].

4.4.2 Hot Forging

Hot forging is a type of thermo-mechanical fabrication process, which uses heat and forging force to deform the material that in turn form the desired shape. Forging force is applied uniaxial with the help of press of hammer. Forging is usually applied to fabricate ingots and billets of shape memory alloys [3]. Hot forging is recommended for the shape memory alloys as the material fabricated by hot forging presents finer

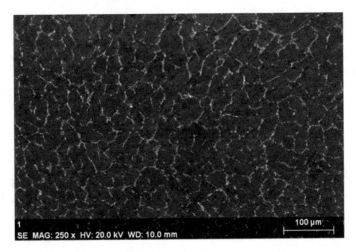

Fig. 4.13 Microstructure of NiTi material (optimum forming condition) fabricated by hot forging [49]

grains due to dynamic recrystallization. Varying orientation of microstructure is decreased with hot forging [48]. Design of die and its material selection for hot forging are most important criteria for shape memory alloys wherein high durability in loading is to be ensured [3, 48]. It is further reported that addition of outer covering foreign material in hot forging of shape memory alloys serves as a jacket and that lead to improving properties of shape memory alloys with little need of jacket removal process and post heat treatment. Subsequent finishing processes such as machining and super finishing are avoided as final product is with better dimensions and surface finish because of jacket covering material [3]. Shape memory alloy of Ni-Mn-Ga is investigated by hot forging. It is exhibited that favorable hot forging parameters can fabricate a complex shape memory alloy part with excellent shape memory effects and without any major defects [48, 49]. Optimum forming condition obtained after hot forging of NiTi shape memory alloy is shown through microstructural evaluation in Fig. 4.13.

4.4.3 Hot Rolling

Hot rolling is a processing technique in which the material is heated and forced to press with rollers. Subsequently, the material undergoes plastic deformation that leads to the change in dimensions such as reduction in thickness and increase in length and width [3]. Hot rolling of shape memory alloys needs important process parameters such as heating temperature, pulling tension applied to the material, type of lubricant, and number of passes to control as they govern shape memory effect and other mechanical properties [3]. It is very important to set the heating temperature. If the temperature is higher than certain level, the shape memory material undergoes dynamic recrystallization or full strain recovery. Further, the dynamic recrystallization and strain recovery remain incomplete because of fast cooling rate as material is initially heated. Issues of work hardening and hardness are faced with improper setting of process parameters for shape memory alloys. Annealing process is recommended for hot rolled shape memory alloys to relive the residual stresses [3, 50].

Hot compression deformation is another type of thermo mechanical processing as reported for NiTi [51]. It is reported that larger equiaxed grains are obtained with increasing the deformation temperatures or decreasing the stain rates. The dynamic recrystallization of NiTi is greatly influenced by the degree of deformation that is consequently affected by stain rate. Critical degree of deformation needs to be identified as larger deformation beyond limit leads to the finer equiaxed grains [51].

4.5 Mechanical Processing

Mechanical processing is a category in which the processing is done by mechanical loading that causes material deformation and leads to the change in properties. Mechanical processing such as equal channel angular extrusion, cold forging, and cold drawing attempted for shape memory alloys are discussed here as under.

4.5.1 Equal Channel Angular Extrusion

Equal channel angular extrusion (ECAE) is a process in which sever plastic deformation is caused by high axial pressure as shown in Fig. 4.14. The large amount of uniform strain is applied and the cross-section remained unchanged while elongated refined grains are obtained from equiaxed large grains that finally results in billets. ECAE is a better process in terms of uniform microstructure, control over grain morphology and texture, and easy processing with respect to the other processes. ECAE is challenging for shape memory alloys as they have high flow strength in austenite along with limited ductility. In case of ECAE of shape memory alloys, the design of pressing tool is required to be high end to handle high strength levels that can limit the friction. Transformation behavior and microstructural changes of NiTi material are greatly affected by ECAE process [52].

4.5.2 Cold Forging

Cold forging is a process in which the material is not heated and mechanically deformed with the help of adequate forging force. The strain rate is very high that in

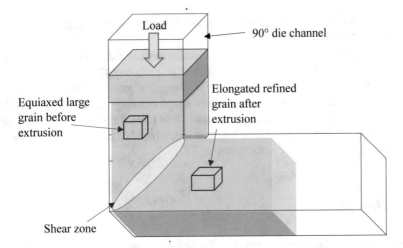

Fig. 4.14 Equal channel angular extrusion

turn causes unstable deformation. Therefore, processing of shape memory alloys by cold rolling causes lower quality in terms of formation of defects and surface finish [3]. It is reported that inter-pass annealing is recommended frequently to enhance the quality of cold-forged shape memory alloy [53]. However, frequent inter-pass annealing with cold forging at low strain deformation can lead to low work hardening that subsequently increases the possibility of defect formation. Further, it is suggested that cold forging of shape memory alloys can be performed with small oxide layer that absorbs applied load and protect underneath material [3].

4.5.3 Cold Rolling and Cold Drawing

Cold rolling is a process in which material subjected from two rollers that provide compressive forces to the material. Shape memory alloys are processed by cold rolling processes with subsequent annealing processing in order to tailor fine grain microstructure. A large number of investigations are available for cold rolling of shape memory alloys [54–56]. It is reported that cold rolled shape memory alloys (especially NiTi material) and subsequent annealing lead to nanostructured microstructure, which presents an excellent shape memory effect along with enhanced mechanical properties [3]. In case of cold rolling of shape memory alloys, factors such as thickness reduction during each pass, pulling tension magnitude, and lubrication are reported as most influenced parameters that affects microstructure and properties [3, 54–56]. Optimization of these parameters is mandatory to have a shape memory effect and desired properties [3]. Sometimes, the oxide layers are kept on the shape memory alloys that act as a lubricant. However, thick oxide layer can cause defects on the surface of the shape memory alloys. Shape memory alloys processed with thick oxide layer also causes problems in shape memory effects and properties [54–56].

Cold drawing is a similar type of process like extrusion and cold rolling wherein shape memory alloy rod or wire is subjected to a die to pull the material pull the material that subsequently resulting in an elongated grains by having decreased dimensions [3]. In case of cold drawing of shape memory alloys, the intermediate annealing is needed as the plastic deformation is in large form similar to cold rolling. The use of lubricants or protective layers is recommended for shape memory alloys that enhances the finishing and processing [57].

4.6 Summary

Different processing techniques including powder metallurgy, additive processing, mechanical processing, and thermo-mechanical processing attempted for various shape memory alloys are considered and discussed for process principle, process condition, process factors, microstructures, properties, and applications. Optimum processing parameters and post-processing are recommended in majority of the processing techniques considering properties of shape memory alloys. NiTi shape memory alloy is most investigated material observed considering wide range of applications and its popularity. Further research and development is required to be done to establish the field further.

References

1. M.H. Elahinia, M. Hashemi, M. Tabesh, S.B. Bhaduri, Manufacturing and processing of NiTi implants: a review. Prog. Mater. Sci. **57**(5), 911–946 (2012)
2. M. Elahinia, N.S. Moghaddam, M.T. Andani, A. Amerinatanzi, B.A. Bimber, R.F. Hamilton, Fabrication of NiTi through additive manufacturing: a review. Prog. Mater. Sci. **83**, 630–663 (2016)
3. A.P. Markopoulos, I.S. Pressas, D.E. Manolakos, Manufacturing processes of shape memory alloys, in *Materials Forming and Machining: Research and Development* (2015), p. 155
4. S.M. Tang, C.Y. Chung, W.G. Liu, Preparation of CuAlNi-based shape memory alloys by mechanical alloying and powder metallurgy method. J. Mater. Process. Technol. **63**(1–3), 307–312 (1997)
5. Z. Li, Z.Y. Pan, N. Tang, Y.B. Jiang, N. Liu, M. Fang, F. Zheng, Cu–Al–Ni–Mn shape memory alloy processed by mechanical alloying and powder metallurgy. Mater. Sci. Eng., A **417**(1), 225–229 (2006)
6. R.D. Jean, T.Y. Wu, S.S. Leu, The effect of powder metallurgy on Cu-Al-Ni shape memory alloys. Scr. Metall. Mater. **25**(4), 883–888 (1991)
7. L.G. Bujoreanu, S. Stanciu, B. Özkal, R.I. Comăneci, M. Meyer, Comparative study of the structures of Fe-Mn-Si-Cr-Ni shape memory alloys obtained by classical and by powder metallurgy, respectively, in *European Symposium on Martensitic Transformations* (EDP Sciences, 2009), p. 05003
8. B. Bertheville, J.E. Bidaux, Alternative powder metallurgical processing of Ti-rich NiTi shape-memory alloys. Scripta Mater. **52**(6), 507–512 (2005)

9. M. Bram, A. Ahmad-Khanlou, A. Heckmann, B. Fuchs, H.P. Buchkremer, D. Stöver, Powder metallurgical fabrication processes for NiTi shape memory alloy parts. Mater. Sci. Eng., A **337**(1), 254–263 (2002)

10. M. Köhl, T. Habijan, M. Bram, H.P. Buchkremer, D. Stöver, M. Köller, Powder metallurgical near-net-shape fabrication of porous NiTi shape memory alloys for use as long-term implants by the combination of the metal injection molding process with the space-holder technique. Adv. Eng. Mater. **11**(12), 959–968 (2009)

11. B. Yuan, C.Y. Chung, M. Zhu, Microstructure and martensitic transformation behavior of porous NiTi shape memory alloy prepared by hot isostatic pressing processing. Mater. Sci. Eng., A **382**(1), 181–187 (2004)

12. N. Zhang, P.B. Khosrovabadi, J.H. Lindenhovius, B.H. Kolster, TiNi shape memory alloys prepared by normal sintering. Mater. Sci. Eng., A **150**(2), 263–270 (1992)

13. S.M. Green, D.M. Grant, N.R. Kelly, Powder metallurgical processing of Ni–Ti shape memory alloy. Powder Metall. **40**(1), 43–47 (1997)

14. S. Wu, C.Y. Chung, X. Liu, P.K. Chu, J.P.Y. Ho, C.L. Chu, Y.L. Chan, K.W.K. Yeung, W.W. Lu, K.M.C. Cheung, K.D.K. Luk, Pore formation mechanism and characterization of porous NiTi shape memory alloys synthesized by capsule-free hot isostatic pressing. Acta Mater. **55**(10), 3437–3451 (2007)

15. M.D. McNeese, D.C. Lagoudas, T.C. Pollock, Processing of TiNi from elemental powders by hot isostatic pressing. Mater. Sci. Eng., A **280**(2), 334–348 (2000)

16. B. Yuan, C.Y. Chung, X.P. Zhang, M.Q. Zeng, M. Zhu, Control of porosity and superelasticity of porous NiTi shape memory alloys prepared by hot isostatic pressing. Smart Mater. Struct. **14**(5), S201 (2005)

17. E. Schüller, O.A. Hamed, M. Bram, D. Sebold, H.P. Buchkremer, D. Stöver, Hot isostatic pressing (HIP) of elemental powder mixtures and prealloyed powder for NiTi shape memory parts. Adv. Eng. Mater. **5**(12), 918–924 (2003)

18. S.L. Wu, X.M. Liu, P.K. Chu, C.Y. Chung, C.L. Chu, K.W.K. Yeung, Phase transformation behavior of porous NiTi alloys fabricated by capsule-free hot isostatic pressing. J. Alloy. Compd. **449**(1), 139–143 (2008)

19. L. Krone, E. Schüller, M. Bram, O. Hamed, H.P. Buchkremer, D. Stöver, Mechanical behaviour of NiTi parts prepared by powder metallurgical methods. Mater. Sci. Eng., A **378**(1), 185–190 (2004)

20. L. Krone, J. Mentz, M. Bram, H.P. Buchkremer, D. Stöver, M. Wagner, G. Eggeler, D. Christ, S. Reese, D. Bogdanski, M. Köller, The potential of powder metallurgy for the fabrication of biomaterials on the basis of nickel-titanium: a case study with a staple showing shape memory behaviour. Adv. Eng. Mater. **7**(7), 613–619 (2005)

21. B.Y. Li, L.J. Rong, V.E. Gjunter, Y.Y. Li, Porous Ni-Ti shape memory alloys produced by two different methods. Z. Metall. **91**(4), 291–295 (2000)

22. A. Biswas, Porous NiTi by thermal explosion mode of SHS: processing, mechanism and generation of single phase microstructure. Acta Mater. **53**(5), 1415–1425 (2005)

23. D.C. Dunand, P. Müllner, Size effects on magnetic actuation in Ni-Mn-Ga shape-memory alloys. Adv. Mater. **23**(2), 216–232 (2011)

24. R.S. Kishore, R.K. Nandhakumaar, V. Gokuul, A. Siddharthan, Characterization of mechanochemically synthesized Cu-Al-Mn shape memory alloy powders and spark plasma sintered compacts

25. R.A. Portier, P. Ochin, A. Pasko, G.E. Monastyrsky, A.V. Gilchuk, V.I. Kolomytsev, Y.N. Koval, Spark plasma sintering of Cu–Al–Ni shape memory alloy. J. Alloy. Compd. **577**, S472–S477 (2013)

26. C. Shearwood, Y.Q. Fu, L. Yu, K.A. Khor, Spark plasma sintering of TiNi nano-powder. Scripta Mater. **52**(6), 455–460 (2005)

27. K. Ito, W. Ito, R.Y. Umetsu, S. Tajima, H. Kawaura, R. Kainuma, K. Ishida, Metamagnetic shape memory effect in polycrystalline NiCoMnSn alloy fabricated by spark plasma sintering. Scripta Mater. **61**(5), 504–507 (2009)

28. H. Meier, C. Haberland, J. Frenzel, Structural and functional properties of NiTi shape memory alloys produced by selective laser melting, in *Innovative Developments in Design and Manufacturing: Advanced Research in Virtual and Rapid Prototyping* (2011), pp. 291–296

29. B.V. Krishna, S. Bose, A. Bandyopadhyay, Fabrication of porous NiTi shape memory alloy structures using laser engineered net shaping. J. Biomed. Mater. Res. B Appl. Biomater. **89**(2), 481–490 (2009)

30. C. Haberland, H. Meier, J. Frenzel, On the properties of Ni-rich NiTi shape memory parts produced by selective laser melting, in *ASME 2012 Conference on Smart Materials, Adaptive Structures and Intelligent Systems* (American Society of Mechanical Engineers, 2012), pp. 97–104

31. I.V. Shishkovsky, L.T. Volova, M.V. Kuznetsov, Y.G. Morozov, I.P. Parkin, Porous biocompatible implants and tissue scaffolds synthesized by selective laser sintering from Ti and NiTi. J. Mater. Chem. **18**(12), 1309–1317 (2008)

32. T. Niendorf, F. Brenne, P. Krooß, M. Vollmer, J. Günther, D. Schwarze, H. Biermann, Microstructural evolution and functional properties of Fe-Mn-Al-Ni shape memory alloy processed by Selective laser melting. Metall. Mater. Trans. A **47**(6), 2569–2573 (2016)

33. T. Gustmann, A. Neves, U. Kühn, P. Gargarella, C.S. Kiminami, C. Bolfarini, J. Eckert, S. Pauly, Influence of processing parameters on the fabrication of a Cu-Al-Ni-Mn shape-memory alloy by selective laser melting. Addit. Manuf. **11**, 23–31 (2016)

34. E.M. Mazzer, C.S. Kiminami, P. Gargarella, R.D. Cava, L.A. Basilio, C. Bolfarini, W.J. Botta, J. Eckert, T. Gustmann, S. Pauly, Atomization and selective laser melting of a Cu-Al-Ni-Mn shape memory alloy, in *Materials Science Forum*, vol. 802 (2014)

35. P. Gargarella, C.S. Kiminami, E.M. Mazzer, R.D. Cava, L.A. Basilio, C. Bolfarini, W.J. Botta, J. Eckert, T. Gustmann, S. Pauly, Phase formation, thermal stability and mechanical properties of a Cu-Al-Ni-Mn shape memory alloy prepared by selective laser melting. Mater. Res. **18**, 35–38 (2015)

36. I.V. Shishkovsky, M.V. Kuznetsov, Y.G. Morozov, Porous titanium and nitinol implants synthesized by SHS/SLS: microstructural and histomorphological analyses of tissue reactions. Int. J. Self Propag. High Temp. Synth. **19**(2), 157–167 (2010)

37. A. Bandyopadhyay, B. Krishna, W. Xue, S. Bose, Application of laser engineered net shaping (LENS) to manufacture porous and functionally graded structures for load bearing implants. J. Mater. Sci. - Mater. Med. **20**(1), 29 (2009)

38. J. Otubo, O.D. Rigo, C.M. Neto, P.R. Mei, The effects of vacuum induction melting and electron beam melting techniques on the purity of NiTi shape memory alloys. Mater. Sci. Eng., A **438**, 679–682 (2006)

39. J. Otubo, O.D. Rigo, C.M. Neto, M.J. Kaufman, P.R. Mei, Scale up of NiTi shape memory alloy production by EBM. In *Journal de Physique IV (Proceedings)*, vol. 112 (EDP sciences, 2003), pp. 873–876

40. J. Otubo, O.D. Rigo, C.D. Moura Neto, M.J. Kaufman, P.R. Mei, Low carbon content NiTi shape memory alloy produced by electron beam melting. Mater. Res. **7**(2), 263–267 (2004)

41. K. Mehta, Advanced joining and welding techniques: an overview, in *Advanced Manufacturing Technologies* (Springer International Publishing, 2017), pp. 101–136

42. K.P. Mehta, V.J. Badheka, A review on dissimilar friction stir welding of copper to aluminum: process, properties, and variants. Mater. Manuf. Processes **31**(3), 233–254 (2016)

43. Z.Y. Ma, Friction stir processing technology: a review. Metall. Mater. Trans. A **39**(3), 642–658 (2008)

44. A. Barcellona, L. Fratini, D. Palmeri, C. Maletta, M. Brandizzi, Friction stir processing of Niti shape memory alloy: microstructural characterization. Int. J. Mater. Form. **3**, 1047–1050 (2010)

45. B. London, J. Fino, A.R. Pelton, M. Mahoney, T.J. Lienert, T.M.S. Warrendale, Friction stir processing of Nitinol. Friction Stir Weld. Process. **III** (2005)

46. D.R. Ni, J.J. Wang, Z.N. Zhou, Z.Y. Ma, Fabrication and mechanical properties of bulk NiTip/Al composites prepared by friction stir processing. J. Alloy. Compd. **586**, 368–374 (2014)

47. M. Dixit, J.W. Newkirk, R.S. Mishra, Properties of friction stir-processed Al 1100–NiTi composite. Scripta Mater. **56**(6), 541–544 (2007)
48. D.Y. Cong, Y.D. Wang, X. Zhao, L. Zuo, R.L. Peng, P. Zetterström, P.K. Liaw, Crystal structures and textures in the hot-forged Ni-Mn-Ga shape memory alloys. Metall. Mater. Trans. A **37**(5), 1397–1403 (2006)
49. J.T. Yeom, J.H. Kim, J.K. Hong, S.W. Kim, C.H. Park, T.H. Nam, K.Y. Lee, Hot forging design of as-cast NiTi shape memory alloy. Mater. Res. Bull. **58**, 234–238 (2014)
50. F.B. Fernandes, K.K. Mahesh, A. dos Santos Paula, Thermomechanical treatments for Ni-Ti alloys, in *Shape Memory Alloys-Processing, Characterization and Applications* (InTech, 2013)
51. S.Y. Jiang, Y.Q. Zhang, Y.N. Zhao, Dynamic recovery and dynamic recrystallization of NiTi shape memory alloy under hot compression deformation. Trans. Nonferrous Met. Soc. China **23**(1), 140–147 (2013)
52. I. Karaman, A.V. Kulkarni, Z.P. Luo, Transformation behaviour and unusual twinning in a NiTi shape memory alloy ausformed using equal channel angular extrusion. Phil. Mag. **85**(16), 1729–1745 (2005)
53. M.H. Wu, Fabrication of nitinol materials and components, in *Materials Science Forum*, vol. 394 (Trans Tech Publications, 2002), pp. 285–292
54. V. Demers, V. Brailovski, S.D. Prokoshkin, K.E. Inaekyan, Optimization of the cold rolling processing for continuous manufacturing of nanostructured Ti–Ni shape memory alloys. J. Mater. Process. Technol. **209**(6), 3096–3105 (2009)
55. Y. Facchinello, V. Brailovski, S.D. Prokoshkin, T. Georges, S.M. Dubinskiy, Manufacturing of nanostructured Ti–Ni shape memory alloys by means of cold/warm rolling and annealing thermal treatment. J. Mater. Process. Technol. **212**(11), 2294–2304 (2012)
56. K. Gall, J. Tyber, G. Wilkesanders, S.W. Robertson, R.O. Ritchie, H.J. Maier, Effect of microstructure on the fatigue of hot-rolled and cold-drawn NiTi shape memory alloys. Mater. Sci. Eng., A **486**(1), 389–403 (2008)
57. J. Otubo, P.R. Mei, S. Koshimizu, Production and characterization of stainless steel based Fe-Cr-Ni-Mn-Si (-Co) shape memory alloys. J. Phys. IV **5**(C8), C8–427 (1995)

Index

A
Additive processing, 61, 67–69, 71, 78
Adhesive bonding, 39, 54, 57
Annealing, 47, 48, 76–78

B
Biomedical, 6
Brazing and soldering, 39, 55–57

C
Cold drawing, 76, 78
Cold forging, 76, 77
Cold rolling, 77, 78
Cryogenic, 9, 19–24, 26, 27, 33

D
Diffusion bonding, 39, 43, 57

E
Electro Discharge Machining (EDM), 10, 11, 13
Electron Beam Melting (EBM), 68, 72, 73
Electron Beam Welding (EBW), 39, 40, 46, 47, 57
Energy, 3, 6, 9, 11, 14, 16, 19, 41, 47, 53, 56, 67, 68, 70
Equal channel angular extrusion, 76, 77
Explosive welding, 39, 50, 53, 54, 57

F
Fabrication, 1, 4, 6, 15, 62, 65, 67, 70–74
Friction Stir Processing (FSP), 73, 74
Friction Stir Welding (FSW), 39, 50, 51, 57, 73
Friction welding, 39, 50, 51, 57

H
Heat treatment, 4, 5, 48, 50, 54, 72, 73, 75
Hot forging, 74, 75
Hot isostatic pressing, 61, 63, 64
Hot rolling, 73, 75

I
InterMetallic Compounds (IMCs), 40, 61

L
Laser Beam Machining (LBM), 14, 15
Laser Beam Welding (LBW), 39, 40, 43, 46, 57
Laser Engineered Net Shaping (LENS), 68, 71

M
Machinability, 9, 10, 13, 18–22, 26, 28, 30, 32
Mechanical processing, 61, 73, 76
MEMS, 1, 71
Metal injection molding, 61, 65, 66
Microstructure, 3, 5, 6, 27, 28, 45, 47, 48, 52, 56, 57, 63, 68, 71, 74–78
Minimum Quantity Lubrication (MQL), 9, 19, 21

P
Phase transformation, 1, 3, 4, 22, 47, 48, 50, 52, 57, 61, 64, 71
Plasma welding, 39, 42, 57
Powder metallurgy, 4, 6, 61, 63, 65–67, 78
Processing, 1, 6, 14, 17, 54, 61, 66, 67, 72, 75, 76, 78
Process parameters, 18, 44, 46, 51, 53, 54, 61, 66, 69, 71, 74–76

© The Author(s), under exclusive licence to Springer Nature Switzerland AG 2019
K. Mehta and K. Gupta, *Fabrication and Processing of Shape Memory Alloys*,
Manufacturing and Surface Engineering, https://doi.org/10.1007/978-3-319-99307-2

Printed in the United States
By Bookmasters